The Cosmic Onion

Quarks and the nature of the universe

The subatomic layers of the cosmic onion

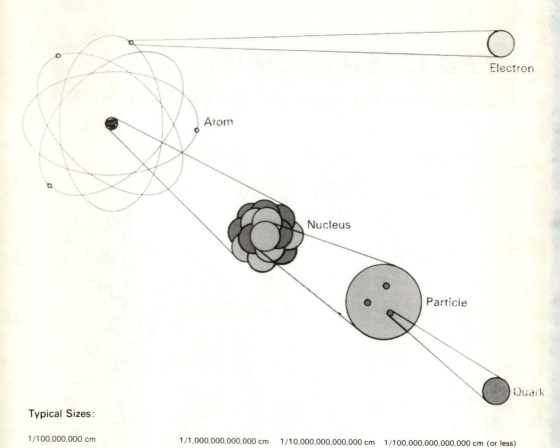

Electron

Atom

Nucleus

Particle

Quark

Typical Sizes:

1/100,000,000 cm   1/1,000,000,000,000 cm   1/10,000,000,000,000 cm   1/100,000,000,000,000 cm (or less)

These minute distances can be more easily written as 10^{-8} cm for atoms, 10^{-12} cm for nucleus, and 10^{-13} cm for a nuclear particle. Electrons and quarks are the varieties of matter that exist as distances of less than 10^{-14} cm. These are the shortest distances that present technology can probe. (*Photo CERN*)

The Cosmic Onion

Quarks and the nature of the universe

Frank Close

**Senior Principal Scientific Officer,
Rutherford Appleton Laboratory, Oxfordshire**

 AMERICAN INSTITUTE OF PHYSICS

© Frank Close 1983
All rights reserved.
First published in 1983 by Heinemann Educational Books Ltd.;
  reprinted 1984, 1985
Fourth printing, 1986
Printed in the United States of America

Library of Congress Cataloging in Publication Data

Close, F. E.
  The cosmic onion.

  Bibliography: p.
  Includes index.
  1. Quarks. 2. Cosmology. I. Title.
QC793.5.Q252C57 1986   539.7'216   86-3646
ISBN 0-88318-491-5

**In the beginning was the Grim End.
For all who were there.**

Contents

Preface

When I give popular talks about the forces of nature, atoms, quarks, and the latest theories of the Big Bang, I am frequently asked two questions: 'Where can I read more about this?' and 'Why don't you write about it for a larger audience?'.

It is not easy to answer the first question, so I have answered both questions by writing this book. Weinberg's superb book *The First Three Minutes* describes nature's development between 1/100 second and 3 minutes after the Big Bang. During this period the neutrons, protons, nuclei, and atoms which ultimately led to our birth some ten to twenty billion years later were fused; but the discovery of the laws that govern the build-up of the present universe from these elementary constituents are not described there. There are several books in this latter area, but these have now largely been overtaken by the rapid developments in our understanding during the last few years.

Grand Unified Theories have been proposed which imply that the variety in today's cold universe is a remnant of a beautiful unity present in the heat of the Big Bang. This enables us to contemplate seriously the nature of the universe as early as 10^{-33} seconds after the Big Bang—which is ancient history compared to the earliest time that was understood when Weinberg's book appeared in 1977.

That we arrived at this singular ability to contemplate seriously such things (even if our thoughts eventually turn out to be misguided), is an incredible achievement of the human intellect. Is it possible to describe these revolutionary ideas in a way that is accessible to the general public? Some of the ideas are very new and still not fully understood. I have attempted to describe them in a way appropriate to an audience that has some scientific familiarity—say final year at school—while adding further material in 'boxes' alongside so that a larger readership can be catered for without disturbing the flow of the text.

The first half of the book is historical and could provide a background to a modern physics course or be general reading for prospective science undergraduates. It is of sufficient generality that it could be easily assimilated by any non-scientist who is prepared to think. The later chapters are more detailed and require more careful reading, although they are self contained. Unfortunately, knowledge that I have sweated over for ten years cannot be passed on in a way that enables it to be absorbed in ten hours. Oh that life were so easy! If at first you fail to grasp some point, don't give in—persevere and return later. The first chapter in particular is an overview painted with a broad brush, telling what the book is about. The subsequent chapters build on this and fill out the details. I could well imagine that many will read this first chapter superficially before the rest and then, having completed the book, return and read it again with the experience and familiarity they have gained.

At times I have found great difficulties in getting the manuscript into its final form—I hope that readers find it easier in consequence. I must thank Graham Taylor and Sheila Watson in particular for encouraging my perseverence and Pam Coleman and my wife, Gillian Close, for their many hours of typing. The idea of using quark logos as an educational device was inspired by Alvaro de Rujula. I am deeply indebted to many colleagues for their comments and suggestions, in particular Z. Golab-Meyer, Ali Graham, Jim Hines, Tony Hey, John Lewis, David Morgan, Lucy Purkis, Hector and Marco Rubinstein, and Jacques Weyers who have read various versions of the manuscript in great detail and helped me to converge on this final version.

I must also thank Rudolf Peierls for allowing me to read vol. 9 of Niels Bohr's collected works in advance of publication, and for his firsthand reminiscences of the early days of nuclear physics.

F. E. Close
Oxford
September 1982

1 The nature of the universe

For nearly a century, physicists have actively sought to uncover the ultimate structure of matter, the fundamental particles from which the infinite variety around us is built, and to formulate the laws that bind these constituents together building up the atoms, molecules, stars, and galaxies—the very universe itself. Discoveries in the last few years have lead to a belief that the answers to these questions are close at hand or might have already been found. What is creating so much excitement is that some of these new insights suggest that we may also have uncovered a glimpse of the birth of the universe and have identified the processes that fashioned it.

At the start of the century 92 atomic elements were thought to be nature's elementary building blocks. But then the atom was smashed and its inner structure revealed: electrons apparently orbited around a massive compact nucleus.

Electrons are held in place, remote from the nucleus, by the electromagnetic attraction of opposite charges, electrons being negatively and the atomic nucleus positively charged. A temperature of a few thousand degrees is sufficient to break this electromagnetic attraction and liberate electrons from within atoms. The ease with which electrons can be removed is the source of chemistry, biology, and life. Restricted to these relatively cool conditions the nineteenth century scientist was only aware of chemical activity; the heart of the atom, the nucleus, was hidden from view.

The accidental discovery of natural radioactivity by Becquerel in 1896 provided the tool with which the atom could be smashed and the nucleus revealed. The nucleus was later shown to consist of protons and neutrons. The force that clusters protons and neutrons together and builds the nucleus is so strong that vast energies can be released when its grip is broken. The heat from chemical reactions such as the burning of coal in a fire is trifling compared to the nuclear heat coming from the Sun or from the stars. Their energy outputs are so huge that they are visible in the night sky as we look back across space and time, in some cases receiving their light millions of years after it set out on its journey.

At the root of many of these nuclear processes is the transformation of neutrons into protons, which converts one element into another. Hydrogen and helium are fused together in stars, and the neutron/proton transmutation builds them up into nuclei of heavier elements such as carbon, oxygen, iron, and the other important ingredients from which the Earth, the air and our bodies are formed. Although it has dramatic effects, the force that is responsible for the neutron/proton transmutation is rather feeble. It is over a thousand times weaker than the electromagnetic force and nearly a million times weaker than the strong nuclear force.

These three distinct forces, the electromagnetic force, the strong force, and the weak force, together with gravity, control the behaviour of all bulk matter and of biological, chemical, and nuclear phenomena. The vast differences in their strengths are crucial to our existence. The reason for this disparity has been a long-standing puzzle—but one that may be about to be solved.

Another puzzling feature of the forces is the discriminatory manner in which they act. Gravity is exceptional in that it acts attractively between all particles. Electromagnetic interactions act only on particles carrying electrical charge, and the strong interaction acts only within the nucleus: electrons are remote from the nucleus not least because they are unaffected by the strong force. Mysterious particles called neutrinos are only affected noticeably by the weak force and are in consequence not trapped in atoms. Thus, though neutrinos are crucial to our universe's existence they are irrelevant in everyday chemistry.

The mysteries of the forces and the nature of the fundamental building blocks have been central puzzles in our quest to understand nature. The nineteenth century physicist knew only of atoms, and of gravitational and electromagnetic forces. Nuclear particles ejected by natural radioactive decay revealed the nucleus to us, and we became aware of the strong and weak nuclear forces during the first half of this century. In the 1960s huge machines were built, several miles in length, which could accelerate electrons or protons until they were travelling at nearly the speed of light. These subatomic 'bullets' then smashed into targets of nuclear material, ploughing deep into the neutrons and protons within, enabling us to study them in greater detail than previously. For a fraction of a second they were heated up to temperatures higher than in any star. Neutrons and protons were found to consist of more basic particles called quarks, and the strong nuclear force was seen to be a remnant of a much more powerful force which clusters quarks together to build neutrons and protons (Box 1.1).

Under these very hot conditions, nuclear processes took on different aspects from those exhibited at lower temperatures. The strong force acting on quarks, and the electromagnetic force acting on electrons,

Box 1.1 Nature's forces act on particles of matter, building up the bulk material for the universe

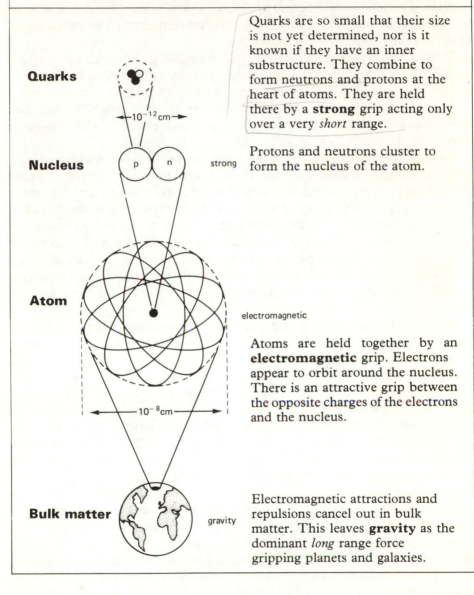

Quarks

Quarks are so small that their size is not yet determined, nor is it known if they have an inner substructure. They combine to form neutrons and protons at the heart of atoms. They are held there by a **strong** grip acting only over a very *short* range.

$\leftarrow 10^{-12}\,cm \rightarrow$

Nucleus

p   n

strong

Protons and neutrons cluster to form the nucleus of the atom.

Atom

electromagnetic

$\leftarrow 10^{-8}\,cm \rightarrow$

Atoms are held together by an **electromagnetic** grip. Electrons appear to orbit around the nucleus. There is an attractive grip between the opposite charges of the electrons and the nucleus.

Bulk matter

gravity

Electromagnetic attractions and repulsions cancel out in bulk matter. This leaves **gravity** as the dominant *long* range force gripping planets and galaxies.

began to appear less dissimilar than hitherto. Radioactivity and electro-magnestism seemed to be two different manifestations of a single force. Quarks, electrons, and neutrinos started to appear as siblings rather than unrelated varieties of elementary particle. The assymetry and disparity of our cold, everyday world seem to be frozen remnants of symmetry and unity prevalent at ultra-high energies.

This discovery has recently had a dramatic psychological effect in science. The disciplines of high-energy physics (the study of subatomic phenomena) and cosmology or astrophysics, seemed far removed—the studies of the ultra-small and the far reaches of outer space. All this has now changed, and the point of union is the physics of the Big Bang (Box 1.2).

Cosmologists now agree that out universe most probably started in a hot Big Bang, beginning its life as a compact fireball where extreme conditions of energy abounded. Collisions of high-energy particles create in a small region of space conditions that are feeble copies of that first Big Bang. But the highest man-made energies are far removed from the intensities that occurred then. The early universe has been likened to the 'poor man's high-energy physics laboratory' and, it is argued, an elegance and beauty which existed then became obscured as the universe cooled. Our high-energy physics experiments have revealed a glimpse of that early unity.

Tempted by that glimpse, physicists have recently constructed Grand Unified Theories, which predict that strong, weak, and electromagnetic forces were originally one, and developed their separate identities as the universe expanded and cooled. Our cold experiences give a poor insight into the processes at work in the Big Bang—the grand unified theories could be the breakthrough that will enable us to understand genesis as never before. But to make this science and not metaphysics we must test the theories. One way is to push experiments to the highest energies possible and see ever clearer visions of the Big Bang extremes. This will not be possible much longer, barring technological breakthroughs, because of the cost of building ever more powerful machines. However, there is still much that can be done in the next decade that should give us many indications of the truth or falsity of the grand unified theories. The complementary approach is to look for exotic, rare phenomena under cold conditions. Processes common in the Big Bang are not entirely absent in cold conditions—they are merely slowed down and so rarer in occurrence.

Two examples will suffice here. It is believed that the strengths of the weak interaction and the electromagnetic interaction were comparable in the heat of the Big Bang. The grand unified theories predict that this remained true during the cooling, all the way down to temperatures

Box 1.2 The 3° background radiation and the Big Bang

In 1964 two astronomers, A. A. Penzias and R. W. Wilson, were using a radio antenna in Holmdel, New Jersey to pick up radio waves coming from outer space. To their annoyance they found a background noise similar to the 'static' that can interfere with the reception of a concert broadcast. This noise was constant, unvarying, and seemingly coming from all directions. At first it was thought that pigeon droppings were the cause, but after cleaning the antenna the noise continued.

Hot bodies glow white, less hot ones yellow or red. Human body heat yields infra-red radiation that shows up in infra-red photography. Cooler bodies emit radio waves (see Box 2.2, Electromagnetic radiation). The background noise was a radio signal corresponding to a temperature only 3° above absolute zero. The source was everywhere, pervading the whole universe.

As a result of this discovery, we now view the universe as having these constituents: matter such as us, the Earth, Sun and stars; radiation from the stars; and this cold 'background radiation'. We believe that the universe erupted in a big bang, spewing matter and radiation outwards from a hot fireball. This is the source of the expanding universe observed today. As the universe expanded, the radiation cooled. The 3° background radiation, the 'static' discovered by Penzias and Wilson, is the cold remnant of that once-hot emission. It is a relic of the Big Bang.

The discovery of this radiation was central in helping to establish the Big Bang theory of genesis. The measurement of its temperature gives a yardstick from which the heat of the early universe can be established. This heat can be compared to the localised heat produced when nuclear particles are collided in the laboratory. We are now beginning to realise that such high-energy nuclear collisions can teach us about the formation and development of the early universe.

which are just within the reach of present high-energy physics accelerators. Below this temperature the weak effects are rare. Note that I say rare, and not entirely absent. We are already aware of weak phenomena by their occasional occurrence and if the grand unified theories are correct their effects should become more noticeable in experiments at higher energies.

Phenomena such as the decay of protons are predicted by grand unified theories to have been present in the heat of the Big Bang. If that was the case, then such decays should still happen occasionally today. The predictions are that in a tank containing about 10^{30} protons, one proton will decay in a year. So you sit and wait and hope that you're looking when the decay occurs. This is an example of the new breed of physics experiments. Grand unified theories predict that exotic phenomena occur at extreme energies and at lower energies they are feeble *but not totally absent*. It is by looking for extremely rare phenomena, frozen remnants of behaviour common during the Big Bang, that tests of the grand unified theories seem most likely to be made in the future.

Thus do we find ourselves at an exciting period in the development of our understanding of natural phenomena. The road that brought us here and the possible future developments are the main themes of this book.

In the early chapters we will meet the history and discoveries that led to the identification of microscopic matter, the basic building blocks, and the forces that cluster them to build the large-scale universe in which we live. The later chapters will describe the new insights that have caused us to recognise the unity in nature under extreme conditions such as prevailed in the early universe. We will end with a brief description of the Grand Unified Theory, its consequences and tests, and its impact on cosmology.

2 Atoms

Over 2500 years ago in Greece, Thales realised that any substance could be classified as either solid, liquid, or gas. This might seem rather obvious to us today as we look back with two millenia of accumulated wisdom, but at that time it was an insight of genius. In addition he went further. Water can exist in each of these forms—might it be the case then that *all* matter is nothing more than water? Although he was wrong in the details, we may credit him with the first suggestion that there is an underlying simplicity responsible for the infinite variety of matter about us.

This article of faith waxed and waned over the centuries. Thales' followers extended the idea by postulating that matter is made from four elements: earth, fire, air, and water. Their manifestations as wind, rain, and sun are still referred to as 'The Elements' though it is with the connotation 'fundamental or elementary building block' that the word 'element' enters the scientific story.

Democritus (585 BC) suggested that matter was built from small *indivisible* particles—in Greek ατομοσ—*atoms*. Combining the ideas of Thales and Democritus would have given the seeds of modern scientific thought, namely that the world is built from a few basic indivisible constituents. However, their ideas lay dormant for 2000 years.

In 1808 the English chemist, John Dalton, proposed the modern atomic theory of matter. Chemistry was becoming a quantitative science and it had been noticed that a wide variety of substances could be formed by combining different quantities of a few elements such as hydrogen, carbon, oxygen, sodium, chlorine and so forth. Over 90 of these elementary substances are now known.

Dalton realised that the ways these elements combined together to form the various substances could be understood if each element was made from atoms. Combining atoms of various elements together made molecules of non-elementary substances. Furthermore he believed that atoms were indivisible; indeed it was for this reason that he honoured Democritus and named them 'atoms'.

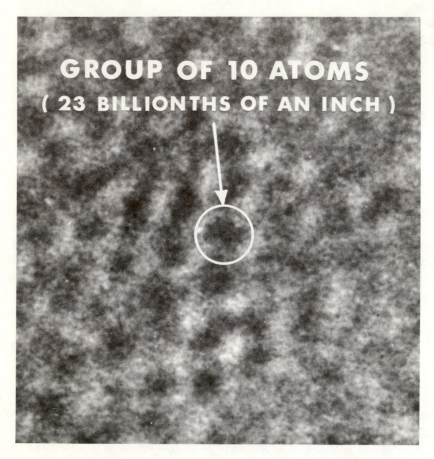

**GROUP OF 10 ATOMS
(23 BILLIONTHS OF AN INCH)**

Photo 2A *Carbon atoms* When viewed through a powerful microscope the granular structure of matter at the atomic and molecular level is revealed. Clustering of carbon atoms is clearly visible in the periodic dark areas of the photograph. Closer examination enables individual atoms of carbon to be resolved. (*Courtesy of Bell Laboratories*)

By the beginning of the present century it was becoming increasingly clear that atoms were not the most fundamental entities in nature. The sequence of events that led to the unravelling of the inner structure of atoms is well known but for future reference, some features are worth mentioning here.

The first indirect hints emerged around 1869 when Dmitri Mendeleev discovered that listing the atomic elements from the lightest— hydrogen—up to the heaviest then known—uranium—caused elements with similar properties to recur at regular intervals (Box 2.1). If each

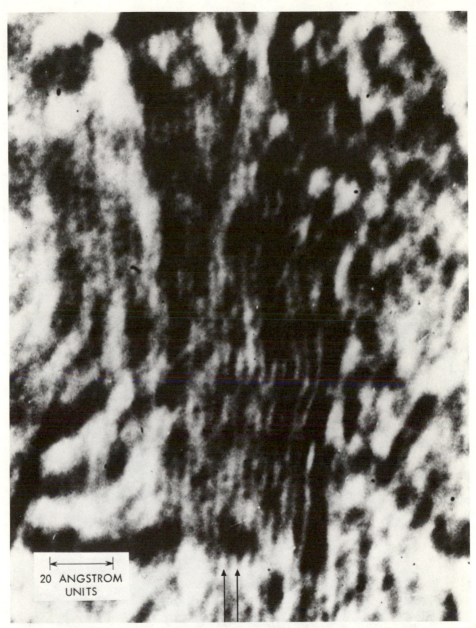

20 ANGSTROM
UNITS

Photo 2B *Layers of carbon atoms* The vertical lines in the centre of this electron micrograph are lattices of carbon atoms viewed from the edge of the crystal plane. Each layer is about one atom thick but individual atoms are not discernable as they overlap one another throughout the plane. This layer structure is the reason why graphite is easily sheared in a plane (along the layer) but is very strong in the perpendicular direction. An angstrom is one hundred-millionth of a centimetre. (*Courtesy of Bell Laboratories*)

Box 2.1   Mendeleev's periodic table of the atomic elements

Tabelle I.

der chemischen Elemente.

			K = 39	Rb = 85	Cs = 133	—	—
			Ca = 40	Sr = 87	Ba = 137	—	—
			★	?Yt = 88?	?Di = 138?	Er = 178?	—
			Ti = 48?	Zr = 90	Ce = 140?	?La = 180?	Th = 231
			V = 51	Nb = 94	—	Ta = 182	—
			Cr = 52	Mo = 96	—	W = 184	U = 240
			Mn = 55	—	—	—	—
			Fe = 56	Ru = 104	—	Os = 195?	—
			Co = 59	Rh = 104	—	Ir = 197	—
			Ni = 59	Pd = 106	—	Pt = 198?	—
Typische Elemente							
H = 1	Li = 7	Na = 23	Cu = 63	Ag = 108	—	Au = 199?	—
	Be = 9,4	Mg = 24	Zn = 65	Cd = 112	—	Hg = 200	—
	B = 11	Al = 27,3	★	In = 113	—	Tl = 204	—
	C = 12	Si = 28	★	Sn = 118	—	Pb = 207	—
	N = 14	P = 31	As = 75	Sb = 122	—	Bi = 208	—
	O = 16	S = 32	Se = 78	Te = 125?	—	—	—
	F = 19	Cl = 35,5	Br = 80	J = 127	—	—	—

The periodic table of the atomic elements, showing the elements known to Mendeleev in 1869 and its modern version. Scandium, gallium, and germanium were successfully predicted by Mendeleev to fill the sites denoted by ★. (*Photos Science Museum, London.*)

element was truly independent of all the others then similarities between them would be merely coincidental and occur randomly. The regularity was indeed striking, and today is understood as arising from the fact that atoms are not elementary, but are instead complex systems of electrons surrounding a compact nucleus, which are held together by the electro-magnetic attraction of opposite charges—electrons being negatively, and the nucleus positively, charged. The experiments that led to this picture of atomic structure were made by Rutherford and his co-workers little more than 70 years ago and were the fruits of a sequence of discoveries spanning several years.

The birth of modern physics is probably marked by Roentgen's discovery of X-rays in 1895. These are now known to be light of short wavelength or high energy produced by atoms in much the same way as is visible light (see Box 2.2), but in 1895 their nature was a mystery. X-rays are most familiar to us today through their medical applications, but the impetus they gave to the development of atomic physics was of great significance. Scientists wanted to understand their nature and in the process, Becquerel in 1896 discovered something totally unexpected. He thought X-rays might be emitted by fluorescent materials—that is, materials which glow in the dark after first being exposed to light. In the course of this investigation he discovered that unexposed potassium-uranium-sulphate emitted radiation which could penetrate black paper and fog a photographic film.

Becquerel's discovery is an illustration of how science does not always proceed in ordered steps. The structure of the atom was not then known (indeed it was another 16 years before Rutherford disentangled its electron-orbiting-nucleus nature) and the source of Becquerel's radiation was still half a century away from being understood. The *nuclei* of the uranium atoms in his potassium-uranium-sulphate were spontaneously breaking up, producing the radiations and a lot of energy—a property that eventually led to the development of the atom bomb and nuclear power.*

Rutherford, while working with Thomson at Cambridge between 1896 and 1900, showed that Becquerel's radiation contained three distinct components, which he named α, β, and γ. The gamma (γ) radiation turned out to be electromagnetic radiation of extremely high energies, much higher even than X-rays. The beta (β) radiation consisted of particles which were soon shown to be negatively charged electrons and the alpha (α) particles were massive positively charged entities—now known to be the nuclei of helium atoms. Having isolated these three

---

* The story of uranium, and the chain of discovery leading to the atomic age is well documented in Bickel's *The Deadly Element* see *Suggestions for Further Reading*.

Box 2.2 Electromagnetic radiation

Jiggle a stick from side to side on the surface of a still pond, and a wave will spread out. A cork floating some distance away will start wobbling when the wave reaches it. Energy has been transferred from the stick to the cork. This energy has been carried by the wave.

If an electric charge is accelerated or jiggled, an 'electromagnetic wave' is transmitted through space. A charge some distance away will be set in motion when the wave arrives. The electromagnetic wave has transported energy from the source to the receiver. A familiar example is an oscillating charge in a radio transmitter. This generates an electromagnetic wave, which transports energy to the charges in your radio aerial.

The speed that water waves travel depends on the distance between successive peaks and troughs (the wavelength). In contrast all electromagnetic waves travel at the same speed—the speed of light. The only difference between light and radio waves is one of wavelength—

The electromagnetic spectrum

Wavelength cm	Example	Use/manifestation
10^{-12}	Photons in	Short wavelengths resolve minute structures in molecules and subatomic phenomena. Also used in biological research
10^{-11}	accelerators	
10^{-10}	γ-rays	
10^{-9}		
10^{-8}	X-rays	X-ray photography. Penetrates skin. Stopped by hard tissue such as bone
10^{-7}		
10^{-6}	Ultra violet	Sun tan
10^{-5}		
10^{-4}	Visible Spectrum ⎧ Blue / Green / Yellow / Orange / Red ⎭	Eye
10^{-3}	Infra-red	Heat. Infrared photography.
10^{-2}		
10^{-1}		
1	Microwaves	Microwave cooking cosmic background
10	UHF	
100	VHF	Radio and TV broadcasting
10^3	TV, FM radio	
10^4	Short wave	
10^5 (1000 metre)	Medium wave (AM)	
1500 metres	BBC Radio 4, Long wave	

Box 2.2—*continued*

radio wavelengths are several metres, light only about a hundred thousandth of a centimetre. The peaks and troughs of a short wave pass a point more frequently than do those of a long wave. It is these oscillations that transport the energy—thus for two waves with the same amplitude more energy is carried by the high frequency (short wavelength) radiation than by the low frequency (long wavelength). Consequently radio waves have low energy, visible light higher energy, X-rays and γ-rays have in turn higher still.

---

components, though not yet knowing how they were formed nor what their true nature was, Rutherford exploited them to study atomic structure.

Electrons were already well known from Thomson's work and are almost two thousand times *lighter* than the lightest atom (hydrogen). They were seen to be emitted by hot metal filaments and also to be ejected from various materials when light struck them. Electrons could be produced from such a wide variety of elements that it was natural to suspect that they probably existed inside all atoms.

In 1902 Rutherford and Soddy discovered that some atoms could spontaneously disintegrate and produce other atoms. In the same decade Pierre Curie and Marie Curie-Sklodowska discovered new radioactive elements—radium and polonium—in the products of uranium's disintegration. Suspicion grew that atoms had an inner structure which differed only slightly between one atom and the next. Small changes in this inner structure would convert one type of atom into another. If electrons were one of the elementary ingredients that built up atoms, what else was there?

Electrons were seen to be deflected by magnetic fields and attracted towards positively charged objects. The rule 'like charges repel, unlike attract' showed that electrons carried negative electrical charge. (We now recognise that the passage of electrical current through matter is due to the flow of these electrons carrying their electrical charge with them.)

If atoms have no net charge themselves, then they must also contain positive electrical charge which precisely balances the electrons' negative charges. When some of the negative charge in one atom is attracted by the positive charge of another, then the two atoms grip one another electromagnetically, binding together to form molecules and macroscopic matter. To understand how this could happen it was first necessary to discover precisely what carried the positive charge in the atom, and this needed some way of looking inside atoms. How could this be done?

If you strike a metal sheet with a hammer you are hitting millions of atoms. The way that the sheet is bent by the blow will reveal how the atoms and molecules are bound together giving the metal its strength, but essentially you learn nothing about individual atoms this way. However, if you hit the metal with something whose size is comparable to an atom, then there is a good chance that an individual atom in the metal will be struck. Thus you may learn something about the atoms themselves, in particular the way that the negative and positive electrical charges are distributed within them.

In the years around 1911 Rutherford and his colleagues at Manchester exploited the alpha particles that were copiously produced in radioactive decay and used them as a tool to hit atoms with. Although their true nature was unknown they have two properties that made them eminently useful for probing inside atoms and for learning about atomic structure. First, alpha particles are emitted from atoms and so are much smaller than them. Second, they have positive electrical charge and so will be repelled by the positive charges in the atoms.*

Alpha particles are several thousand times heavier than electrons and as a result of this great bulk their motion is not significantly disturbed by the electrons. Thomson, the discoverer of the electron, believed that positive charges were spread diffusely throughout an atom, possibly carried by light particles such as the electron. If his idea was correct then the massive alpha-particles would plough straight through the atoms and suffer only a small deflection from their path.

One day in 1909, Ernest Marsden, a student of Rutherford's, reported that when he fired alpha particles at a thin wafer of gold most of them indeed passed straight through but about one in 10 000 bounced back. Can cannon-balls recoil from peas? Rutherford later remarked that it was the most incredible event that had happened in his life: 'It was as if you had fired a 15-inch shell at a piece of tissue paper and it came back and hit you'. Somewhere in the gold atoms must be concentrations of material much more massive than alpha particles.

In 1911 Rutherford announced his solution to the puzzle. He proposed that all of the atom's positive charge and most of its mass are contained in a compact nucleus at the centre. The nucleus occupies only about 10^{-12} of the atomic volume—hence the rarity of the violent collisions—and the electrons are spread diffusely around outside. Rutherford computed how

---

* You might wonder why alpha particles were not suggested as the carriers of the positive charges in atoms. There were two main reasons. First their mass was about four times that of hydrogen and so you couldn't build hydrogen that way. Second, their charge was twice that of the electron (and of opposite sign, of course) and it was felt most natural that the carrier of positive electricity should have the same magnitude of charge as the electron.

frequently alpha particles would be scattered through various angles and how much energy they would lose if a positively charged dense nucleus was responsible for their recoil. During the next two years Marsden and Hans Geiger scattered alpha particles from a variety of substances and confirmed Rutherford's theory of the nuclear atom.

While this left no doubt about where the positive charge is situated there was still a puzzle about the electrons. Rutherford had suggested that these orbited around the central nucleus so that an atom's structure is analogous to the solar system, the essential differences being the overall scale and that there is electromagnetic instead of gravitational attraction. The problem with this idea was that the laws of physics as then known would not allow such an atom to exist: orbiting electrical charges such as electrons should radiate energy and spiral down into the nucleus within a fraction of a second.

Although this would indeed be true in classical, large-scale physical systems, Niels Bohr realised that previous experience might be totally inadequate for dealing with microscopic systems such as atoms. Indeed there were precedents for such caution. In 1900 Max Planck had shown that light is emitted in distinct microscopic 'packets' or 'quanta' of energy known as photons, and in 1905 Einstein showed that light remains in these packets as it travels across space.

The discovery that radiant energy is quantised led Bohr to propose that the energies of electrons in atoms are also quantised: electrons in atoms can have only certain prescribed energies. Restricted to these particular energy states, electrons cannot radiate energy continuously and smoothly spiral inwards, they can only jump from one energy state to another and emit or absorb energy to keep the total amount of energy constant (over long time-scales energy is conserved) (Box 2.3).

In building his theory of energy quanta, Planck had to introduce a quantity known today as 'Planck's constant' traditionally abbreviated to the symbol h (the combination $h/2\pi$ being denoted $\hbar$). Bohr proposed that the permissible energies of electron orbits in atoms were controlled by the same quantity h. He then applied his idea to the simplest atom, hydrogen, whose nucleus is orbited by just one electron. If the electron stayed in an orbit no energy was radiated but if it jumped from a high-energy to a lower energy state then energy would be radiated. Assuming that this radiated energy was converted to light, Bohr calculated the corresponding wavelengths and found that they matched precisely the mysterious spectrum of hydrogen. Planck's quantum theory, applied successfully to radiation by Einstein, had now been applied to matter with equal success by Bohr.

The ways that atomic spectra (Box 2.4) are modified when their parent atoms are in magnetic or electric fields, were explained by hypothesising

Box 2.3   The uncertainty principle

It is not possible to measure both the position and momentum of a particle with arbitrary precision. To observe an electron we may shine light on it and detect the scattered radiation (photons). In the act of scattering photons the electron recoils and alters its momentum. In the macroscopic world this is of no consequence as the momentum of a massive object is not measurably affected, but for atomic and subatomic particles the inability to determine precisely both spatial position and momentum is a fundamental problem.

If the position of a particle is known to be within some distance r of a point, then its momentum must be indeterminate by at least an amount p where

$$p \times r = \hbar$$

and $\hbar$ is a constant of nature known as 'Planck's constant'. Its magnitude is

$$\hbar = 6.6 \times 10^{-22}\,\text{MeV s} \simeq 1 \times 10^{-27}\,\text{erg s} = 1.05 \times 10^{-34}\,\text{J s}$$

(SI units are now regarded as standard but many particle physicists still use ergs, so both have been included here.) Box 4.2 describes the essential role that Planck's constant plays in setting the size of atoms.

A similar uncertainty principle can be applied to time and energy. The uncertainty in energy at a give time implies that energy conservation can be 'violated' over very short time scales—I put 'violated' in quotes because one cannot detect it—this is the nub of the inability to determine energy *precisely* at a given time. Particles can radiate energy (e.g. in the form of photons) in apparent violation of energy conservation, so long as that energy is reabsorbed by other particles within a short space of time. The more that energy is 'violated' the sooner it must be repaid: the more you overdraw on your account the sooner the bank manager is likely to notice but if you pay it back before you are found out, everyone is satisfied. This virtual violation of energy conservation plays an important role in the transmission of forces between particles (see Box 4.5, Yukawa and the pion, p. 40).

that not just the energy but also the angular momentum of the orbiting electron is quantised; being restricted to integer multiples of Planck's constant $\hbar$. Furthermore, the electron was discovered to have an intrinsic spin of its own, magnitude $\frac{1}{2}\hbar$. All of this followed naturally once Bohr's hypothesis of 'allowed orbits' was accepted, but in turn raised the question as to what ordained that electrons should choose these special states.

The essential clue lay with Planck's quantum theory that had been Bohr's inspiration in the first place. Planck and Einstein had shown that

Box 2.4 The hydrogen spectrum

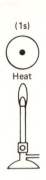
Heat

Electron excited to
the 2s energy level.

$E_\gamma = E_{2s} - E_{1s}$

Electron falls back to
the 1s level and loses
energy. This is emit-
ted as light which
yields a spectral line.

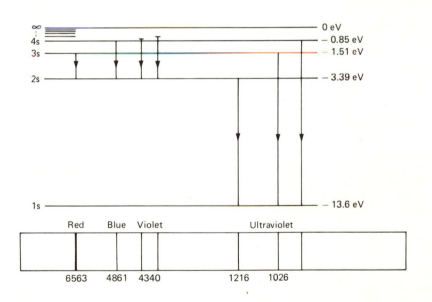

Energy levels in hydrogen for an electron in s states, and some
possible electron jumps that yield spectral light, The numbers denote
the light's wavelength in angstroms: one angstrom is 10^{-8} cm. Visible
light spans the range 3500 to 7000 Angstroms corresponding to energies
of about 3.4 to 1.7 eV.

Box 2.5   The de Broglie waves and allowed orbits in Bohr's atomic model

An electron moving along the path ------ is represented as an electron wave ══════ in de Broglie's theory.

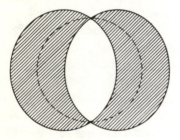

We can imagine a complete wavelength bent into a circle.
When the wave fits the circle precisely this is the first allowed Bohr orbit.

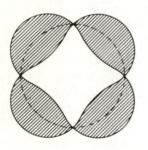

Two wavelengths completing the circle yields the second Bohr orbit which is of higher energy than the first orbit.

Higher energy allowed orbits correspond to larger numbers of wavelengths fitted into the circumference.

radiation with its well known *wavelike* character could act like a staccato burst of *particles* called photons. Twenty years later, in 1925, Louis de Broglie proposed the converse: *particles of matter* can exhibit *wavelike* characteristics. Planck's theory, as generalised by de Broglie, required that the wavelength of a low energy electron would be larger than that of a high energy one. Now imagine an electron circling a nucleus. The lowest energy Bohr orbit ('first orbit') contains exactly one wavelength—as we pass around the Bohr orbit we see the wave peak, then fall to a trough and back to a peak precisely where we started. The second orbit contains exactly two wavelengths, and so on. When the de Broglie electron wave exactly fits into a Bohr orbit the wave reinforces itself by constructive interference and so persists (Box 2.5). On the other hand, if the de Broglie wave does not fit into the orbit the wave interferes destructively with itself and dies out rapidly. Thus discrete orbits and discrete energy states emerge directly from de Broglie's hypothesis on the wave nature of electrons.

This strange marriage of classical ideas with wavelike ingredients was both successful and disturbing. Great efforts were made to understand its workings more deeply, and culminated in the modern quantum theory developed by Schrodinger, Heisenberg, Dirac, and others from 1928 onwards. The history of this is a fascinating story in its own right but this is as far as we need to go in our present context. Modern quantum mechanics gives a more profound description of the atom than we have described but contains within it ideas that correspond to the classical images of 'solid' electrons orbiting and spinning. These conceptually helpful images are widely employed even today and so I shall adhere to this nomenclature in what follows.

3 The nucleus

Neutrons and protons

By the early 1930s the structure of the atom was common knowledge. The rules governing the distribution of the electrons among the atomic energy levels had been successfully worked out, and explained the regularities in chemical properties that Mendeleev had noticed 60 years earlier. The whole of theoretical chemistry was subsumed in the new atomic physics. This caused one of the most prominent theoretical physicists of the time to comment that the basic laws were now formulated and that the prime task was to work out their consequences.

Although the electronic structure of atoms was understood, little was known about the central nucleus beyond the facts that it is very dense and positively charged. In order to learn more about the nucleus, Rutherford and James Chadwick had been bombarding many different nuclei with alpha-particles in a series of experiments during 1919 to 1924.

In the first experiment, nitrogen atoms were the targets. The nitrogen gas filled a tube, into one end alpha-particles were fired and at the far end scattered particles were detected. Rutherford was amazed to find that hydrogen nuclei also emerged even though there had been no hydrogen there to start with! The alpha-particles had ejected hydrogen nuclei out of the nitrogen target. Subsequent experiments used other varieties of nuclei as targets and uniformly showed that nuclei could be transmuted into one another. This proved that nuclei were not featureless dense balls of positive charge but had a detailed internal structure of unknown form.

The simplest atom of all is hydrogen, containing just one electron and a nucleus which is nothing more than a positively charged particle called a proton (the proton is 1836 times more massive than the electron and so provides the bulk of the hydrogen atom's mass). As protons had been ejected from nitrogen and other elements in the experiments described above, Rutherford suggested that it is protons which carry the positive charge of *all* nuclei. The more positive charge that a nucleus carries so the more protons it contains, thus hydrogen has one proton, helium two, oxygen eight and so on. However, on the basis of determination of the

Box 3.1   Neutrons in the nucleus

Why was it necessary to invent the concept of neutrons to make up the nuclear mass—would not pairs of protons and electrons do just as well?

This was indeed one suggestion but it was ruled out by measurements of the nuclear spin. Electrons have spin $\frac{1}{2}\hbar$ (p. 16). Protons have the same amount of spin. It was then discovered that a nitrogen nucleus has an integral value of spin and so must contain an *even* number of spin $\frac{1}{2}\hbar$ particles. It is 14 times heavier than hydrogen and has a net charge of $+7$. The proton-electron model would require 14 protons to yield the mass and 7 electrons to give the net charge. However this is a total of 21 particles—an odd number inconsistent with the even number required by the spin measurement.

If 7 protons and 7 electrons were replaced by 7 neutrons, the net charge and mass are the same as before, but we now have a total of 14 particles. The nuclear spin is satisfactorily described if the neutron has a spin of $\frac{1}{2}\hbar$, identical to the proton.

Proton–electron model	Electrical charge	Proton masses	Number of spin $\frac{1}{2}$ particles
$14\,p^+$	$+14$	14	14
$7\,e^-$	-7	0	7
Total	$+7\checkmark$	14ₓ	21 odd ×

Proton–neutron model	Electrical charge	Nucleon masses	Number of spin $\frac{1}{2}$ particles
$7\,p^+$	$+7$	7	7
$7\,n^0$	0	7	7
Total	$+7\checkmark$	14ₓ	14 even ✓

The nitrogen nucleus has seven times the charge, and fourteen times the mass of a proton. The total spin of the nitrogen nucleus shows that it contains an *even* number of spin $\frac{1}{2}$ constituents. The proton–electron and the proton–neutron model can both fit as regards the charge and mass, but only the proton–neutron model agrees with the even number of constituents.

mass of the chemical elements we know that an oxygen atom is 16 times more massive than a hydrogen atom; its eight protons provide only half of the mass—what contributes the remainder?

Rutherford guessed that there might be another particle as heavy as a proton but with no electrical charge: a 'neutron'. Irene Joliot-Curie, daughter of Marie Curie-Sklodowska, and Frederic Joliot had evidence for the neutron but misinterpreted it. The neutron was discovered in 1932 by Chadwick.

The Joliot-Curies had fired alpha particles at beryllium targets and discovered that an electrically neutral radiation came out which they mistakenly took to be X-rays. Chadwick did the same but in addition placed paraffin wax some distance away from the target. Protons were ejected from the wax when the radiation hit it. Rutherford compared this to H. G. Wells' invisible man—although you could not see him directly, his presence could be detected when he collided with the crowd. Thus it was when the invisible radiation collided with the paraffin wax.

The energy carried by X-rays would easily remove electrons from the atoms in the paraffin but would not eject protons; the proton is so massive that it would merely shudder under the impact. Whatever was knocking the proton out must itself be heavy. Chadwick suggested that here was evidence for a new subatomic particle, similar in mass to the proton but with no electrical charge—the neutron that Rutherford had predicted. Apart from a 1% difference in mass and the presence of electrical charge, the proton and neutron are identical. As they are constituents of the nucleus they are often referred to collectively as 'nucleons' (Box 3.1).

An oxygen nucleus contains 8 protons and 8 neutrons and so is 16 times as massive as hydrogen. All nuclei contain protons and neutrons with the sole exception of the hydrogen nucleus which is a single proton.

In 1932, the same year that Chadwick discovered the neutron, nuclei were split for the first time by *artificial* means. Chadwick and Rutherford's classic experiments had earlier split the nucleus by bombarding it with α-particles, the source of these being *natural* radioactive decays of radium. In contrast, John Cockroft and Ernest Walton used electric fields to accelerate protons to high speed, and then fired them at lithium nuclei. Cockroft and Walton had thus made the first practical nuclear particle accelerator, and created a tool with which more detailed studies of nuclear properties could be made. This was the prototype of the modern particle accelerators that have been used for probing the internal structure of the neutrons and protons themselves (Box 3.2).

Box 3.2   Early particle accelerators

Photo 3A *Cockroft and Walton's original accelerator (1932)* Professor Walton is visible in the counting room. (*Courtesy of Cavendish Laboratory*)

Box 3.2—*continued*

Electrons are electrically charged and so may be accelerated down an evacuated tube if an electric field is applied. In Roentgen's 1895 X-ray tube, electrons were accelerated by a potential of about a thousand volts. When the electrons were deccelerated, e.g. by hitting a screen, electromagnetic radiation was emitted. In modern times this is the principle behind the television.

On lengthening the tube the electrons can be accelerated along its full length to higher energies. Present technology can increase an electron's energy by about 7 MeV for every metre of tube. The Stanford Linear Accelerator (SLAC) in California is 2 miles long and the electrons emerge with energies over 20 GeV. At these energies the electrons can smash into nuclei and even probe the inner structure of protons. The increase in size of electron accelerators over the years is paralleled by proton accelerators.

Rutherford's discovery of the nucleus in 1911 used naturally occurring alpha-particles that had been emitted by radioactive nuclei. One could regard this as a natural accelerator. A controllable source of high energy particles was created by Cockcroft and Walton in 1932 (Photo 3A). Protons were accelerated by a potential of 500 000 volts and were smashed into lithium nuclei, thereby disintegrating them.

To accelerate protons to higher energies required larger distances over which the accelerating force could be applied. Ernest Lawrence accelerated protons around a circular orbit. A magnetic field bent their path around a semicircle, an electric field then gave them a kick, then they were bent round a second semicircle and so on. The protons circled round and round and were accelerated to an energy of 1 MeV. This device is called a 'cyclotron' (Photo 3B).

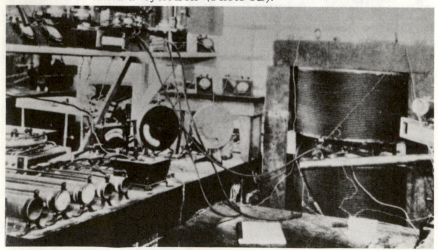

Photo 3B *Lawrence's first cyclotron* The first cyclotron accelerated protons to an energy of 1.2 MeV and was built by E. Lawrence in the USA. It consisted of a large electromagnet with circular pole tips between which protons rotated in a spiral orbit in a vacuum.

Radioactivity and nuclear fission

Protons and neutrons are the common ingredients of *all* nuclei and so one variety of nucleus can transmute into another by absorbing or emitting these particles. The α-particles are themselves helium nuclei that are produced naturally when, for example, uranium nuclei break down into thorium:

$$^{92}U_{238} \rightarrow {}^{90}Th_{234} + {}^{2}He_4$$

The number of protons is shown in the superscript and the number of nucleons (neutrons plus protons) in the subscript. The net numbers of protons and of neutrons are separately conserved throughout: one cluster has broken down into two. This spontaneous decay of nuclei is an example of 'radioactivity' and is the explanation of Becquerel's 1896 discovery of α-particles.

Not all nuclei are radioactive: to all practical purposes most are stable. The most stable nuclei tend to be those where the number of neutrons does not greatly exceed the number of protons. One of the exciting projects in the early 1930s was the bombarding of naturally occurring nuclei with neutrons, in the hope that some neutrons would attach themselves in the target nuclei and form new 'isotopes' (see Box 3.3). Of course, if you fired in neutrons at too high a speed they would shatter the target rather than stick to it. To prevent this Enrico Fermi first slowed the neutrons down by passing them through paraffin and with this technique successfully modified the nuclei of various atoms. He attached neutrons to fluorine, producing a new artificial isotope of that element and did likewise with a total of 42 different nuclear targets.

At last in 1934 he came to the heaviest known element—uranium. Fermi was one of the greatest physicists of all time but at this point he made a mistake for which we may be thankful. After the uranium was irradiated some puzzling phenomena were observed. Fermi assumed these to be evidence that he had produced a new isotope of uranium or had

---

Box 3.3   Isotopes

Every nucleus of a given element contains the same number of protons (Z) but may have different numbers of neutrons (N). Hydrogen usually has one proton and no neutrons but about 0.015% of hydrogen atoms have one additional neutron. This is 'heavy hydrogen' or 'deuterium'. Hydrogen with two neutrons is 'tritium'.

The total number of neutrons and protons is placed as a subscript to the atomic symbol to distinguish the various isotopes. Thus $^{92}U_{235}$ and $^{92}U_{238}$ are two isotopes of uranium: both contain 92 protons but have 143 and 146 neutrons respectively, and hence 235 and 238 nucleons.

Box 3.4 Slow neutrons disintegrating uranium

1 A slow neutron is absorbed by a uranium-235 nucleus.
2 The uranium is now unstable and wobbles like a water drop.
3 The nucleus becomes so deformed that it splits in two.
4 The products are stable nuclei of barium and krypton and two or
  three neutrons. Energy is also released.
5 One of these neutrons might hit another fissionable nucleus of uranium
  so that a chain reaction develops. Energy is released explosively if more
  than one neutron induces another fission.

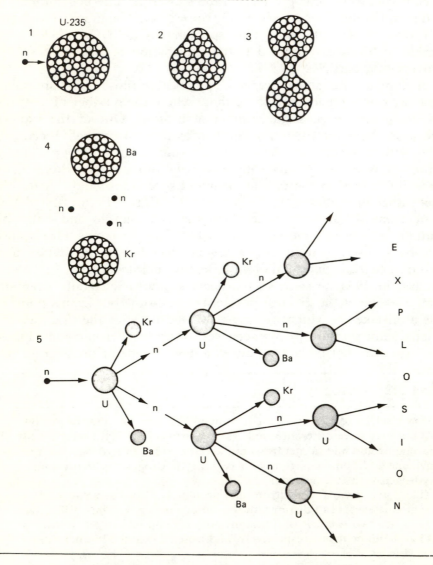

artificially produced the first 'transuranic' element, one place above uranium in Mendeleev's table, unknown on Earth but capable of existence in principle. To have done so would have been a great prize and he was so taken up with this possibility that he missed the real explanation.

Otto Hahn and Lisa Meitner in Germany repeated the experiment and discovered not new elements but barium—an element with a nucleus of 56 protons. This was a great surprise: could slow-moving neutrons so disturb uranium that it split into two almost equal halves? What we now know is that slow neutron bombardment makes a uranium nucleus wobble like a liquid drop and break up. In the fragments there are further slow neutrons that can trigger the break-up of other uranium nuclei, and enormous energies can be released in the ensuing 'chain reaction' (see Box 3.4). The essential ingredients of the so-called atom bomb were unwittingly at hand in Italy and Germany up to five years before World War II.

Lisa Meitner was Jewish and in consequence fled from Germany to Sweden. It was there that Otto Frisch learned of her work with Hahn, came up with the explanation and caused Bohr's famous remark: 'Oh what idiots we have been. It has to be like this!'*

The splitting of nuclei in this way is called nuclear fission. If Fermi had seen the answer in 1934 or if Meitner had not fled from Germany, then the atom bomb might have been developed by Hitler's scientists. As it was, Fermi also fled, and play a major role in the Allies Manhattan project.

Beta-radioactivity and neutrinos

Becqueral discovered three types of radiation in 1896 (Box 3.5): alpha-particles are helium nuclei and gamma rays are extremely energetic photons emitted from the nucleus when the nucleons lose energy in rearrangement (see Box 3.6). Beta radiation consists of electrons—where did they come from?

Heating or irradiating atoms can supply enough energy for one or more of their electrons to be ejected. However, the production of Becquerel's electrons was quite different—they emerged without energy being first supplied to the atom. The source of these is not the electrons in the periphery of the atom, instead they are coming from within the nucleus itself (see Box 3.7).

Isotopes containing a high percentage of neutrons tend to be unstable. They may spontaneously emit alpha-particles and form lighter, more

* O. R. Frisch, *The Interest is focussing on the atomic nucleus* in *Niels Bohr*, ed. S. Rozental, North-Holland 1967, p. 145.

Box 3.5 The three types of nuclear radioactive decay

Alpha and beta decays involve a change in the neutron and proton content of the nucleus and cause it to change into a different species. Following alpha or beta decay the protons and neutrons rearrange themselves and in the process energy is radiated as γ-rays. No disintegration occurs in γ-emission: one or more nucleons is temporarily in a state of high energy ('excited nucleus') which it loses by emitting a γ-ray. The excited nucleus that exists before γ-emission has been labelled with an asterisk.

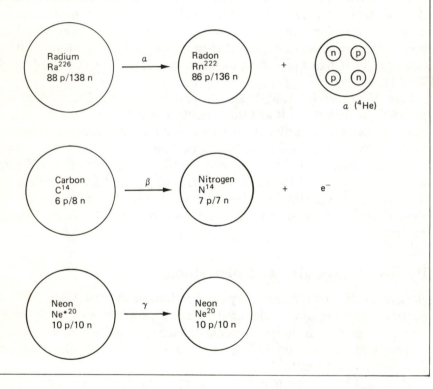

Box 3.6   How photons emerge from atoms

Atoms
Spectra

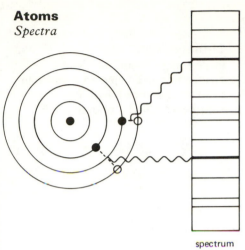

spectrum

In Bohr's model of the atom, light is emitted when an atomic electron jumps from one orbit to another. The energy that the light carries off is equal to the difference of the energies of the electron before and after:

$$E = E_{\text{initial orbit}} - E_{\text{final orbit}}$$

Different energies of light have different colours.

A spectrum gives information on the nature of the energy levels available for the electrons. The colours reveal the differences in energies and from these the actual set of energy levels can be deduced.

X-rays
Heavy atoms have lots of electrons. If an electron drops to the lowest energy level all the way from the highest ones then a lot of energy is emitted as electromagnetic radiation. These energetic bursts of radiation are X-rays (see the section on Roentgen, p. 11).

The nucleus

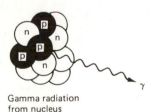

Gamma radiation
from nucleus

The nucleus at the atom's heart has a complicated structure of its own. It contains *neutrons* and *protons*. When protons change from one type of motion to another in the nucleus, very high-energy electromagnetic radiation is sometimes emitted: this is called *gamma rays* (denoted γ). This is analogous to the way that visible light is emitted in atomic rearrangement of electrons.

Box 3.7   How electrons can emerge from atoms

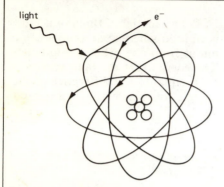

Light
Give *energy* to an *atom* and electrons can be ejected from their orbits. The *energy* may be supplied as *light* ('photoelectric effect')

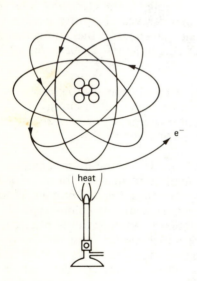

Heat
or *heat* (as in a television tube).

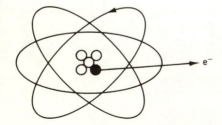

Radioactivity
Radioactivity: a neutron in the *nucleus* converts into a proton and ejects an electron *without energy being supplied*.

stable nuclei, or a neutron can spontaneously convert into a proton and emit an electron in the process. Electrical charge is always conserved in nature and this transmutation is a good example:

$$n^0 \to p^+ + e^-$$

The superscripts denote the electrical charges of n (neutron), p (proton) and e (electron). This neutron decay is known as β-radioactivity and is the source of many nuclear transmutations.

One puzzling feature of this decay was that the proton and electron seemed to have less energy than they ought. According to Einstein's relation $E = mc^2$ the energy released in the decay is

$$E = m_{\text{neutron}}c^2 - [m_{\text{proton}}c^2 + m_{\text{electron}}c^2] = 0.8 \text{ MeV}$$

which should be manifested as kinetic energy of the proton and electron. Energy seemed to have unaccountably disappeared—a phenomenon contrary to centuries of evidence that energy is conserved over long time scales, Box 3.8.

Box 3.8 $E = mc^2$ and particle masses

Energy is conserved over long time scales but can be converted from one form such as potential energy, kinetic energy, chemical energy and heat energy, to another. Einstein showed that another form of energy conversion can occur: energy can be converted into mass and vice-versa. The amount of energy (E) that is produced if mass (m) is destroyed is given by Einstein's famous equation

$$E = mc^2$$

where c is the velocity of light.

The masses of subatomic particles are then given in units of MeV/c^2, more usually shortened to simply MeV. Thus the proton's mass is said to be 939 MeV.

In addition to the imbalance of energy, anomalous behaviour was seen in the directions that the proton and electron moved along after the neutron had decayed (Box 3.9). If a stationary neutron breaks down into two particles then they should move off in opposite directions along a straight line. However, the proton and electron moved off at an angle, as if some unseen third particle was also being produced in the decay. The energy that this particle carries off is given by the difference of the initial neutron's energy and the proton and electron energies, and is the cause of the apparent imbalance in energies that had been noticed.

Box 3.9   The neutrino

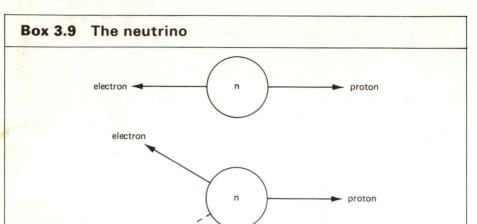

If a neutron decayed to two particles then they would move off along a
straight line as in the first diagram. In practice they are seen to move off
at an angle to one another due to a third (invisible) particle being
produced. This is the *neutrino*.

The particle carries no electrical charge and may have no mass of its
own. Its existence was postulated in 1931 by Wolfgang Pauli in order to
explain the otherwise anomalous energy loss in the neutron decay, but it
was not detected directly until 1956. It was named the 'neutrino' (to
distinguish it from the neutron) and it is conventionally denoted by the
symbol ν.

Neutrinos interact so weakly with other matter that they are very hard
to detect. An illustration of this is the fact that a neutrino produced in
beta decay could pass clean through the Earth as if we were empty space.
Thus neutrinos produced in the sun shine down on us by day and up
through our beds by night. Pauli was so certain that the neutrino could
not be detected that he wagered a case of champagne as incentive.

At the Savannah River nuclear reactor in the USA, over a million
million neutrinos emerge each second per square centimetre, created by
the radioactive decays of the reactor material. This enormous concentration of neutrinos gave the possibility that if many tons of cadmium
solution was placed alongside, then occasionally a neutrino would
interact with it. In 1956, F. Reines and C. L. Cowan observed the
products of the interaction and so inferred the incident neutrino's
existence. (This is similar in spirit to the way that the neutron had
revealed itself by ejecting protons from paraffin wax, p. 22.) Pauli paid
up the champagne that he had promised twenty-five years before.

4 *The forces of nature*

Gravity and electromagnetism

By the 1930s, with the structure of the atom established and the composition of the nucleus under active investigation, there was optimism that the basic building blocks of matter had been isolated: electrons and neutrinos, protons and neutrons. The outstanding theoretical task was to formulate the laws governing their combination and thus explain how matter and the world about us are formed.

Electromagnetic and gravitational forces had been known for a long time. Gravity mutually attracts all matter whereas electromagnetism can attract *or* repel. The familiar adage 'unlike charges attract, like charges repel' is a first step in understanding the structure of the atom; the electrons have negative charge while the nucleus has positive charge and so the binding of the atom is due to the electromagnetic attraction of the electrons and the nucleus.

Electromagnetic forces are intrinsically much more powerful than gravity, so why was gravity identified first? In most large lumps of matter positive and negative charges cancel out, cutting off electromagnetism's sphere of influence and leaving the all-attractive force of gravity as dominant. Gravity is particularly noticeable over astronomical distances. The power of the electromagnetic force is felt in those situations where the positive and negative charges' effects are not precisely cancelled. The most familiar example is where the orbiting and spinning motions of the charges in atoms give rise to readily observable magnetic effects—a small magnet can attract to itself a lump of metal, overcoming the downward pull that the metal is experiencing from the gravitational attraction of the whole earth. On a small scale, such as between individual atoms, the effects of gravity are negligible and it plays no observable role in describing known atomic or nuclear phenomena.

In the 19th century various electric and magnetic effects had been recognised for some time but they were still not well understood. The behaviour of electrical charges in the presence of other electrical charges is described by Coulomb's law, and the force between two wires carrying electrical currents (electrical charges in motion) by Biot and Savart's law.

Box 4.1   The four fundamental forces

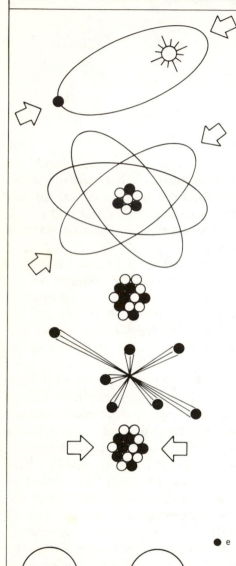

Gravitational force
This acts between all particles. Pulls matter together. Binding force of the solar system and galaxies.

Electromagnetic force
Unlike charges attract. The cloud of negatively charged electrons is held around a positively charged nucleus. Binding force of atoms.

Strong force
The nucleus of an atom contains protons. Like charges repel by the electromagnetic force. But the nucleus doesn't blow apart because the strong force attracts protons and neutrons and binds the nucleus.

This is now believed to be a remnant of a more powerful **colour force** acting on quarks inside the protons and neutrons.

Weak force
This causes radioactive decay of some nuclei. A neutron breaks up, emits an electron and becomes a proton. When operating in this way this force is a thousand million times weaker than the strong nuclear force. At high energies it is not so weak and begins to act in a similar way to the electromagnetic force. At very high energies the electromagnetic and weak forces appear to be intimately related.

e

ν

n

p

before

after

Box 4.2   Nature's natural scales

The gravitational force between two objects whose masses are m and M and which are a distance r apart is

$$F = GMm/r^2$$

If the force F is measured in newtons, m, M in kilograms and r in metres then

$$G = 6.67 \times 10^{-11}\,\mathrm{N\,m^2\,kg^{-2}}$$

G is a fundamental quantity, the gravitational constant, that has dimensions. Similarly the velocity of light ($c = 3 \times 10^8\,\mathrm{m\,s^{-1}}$) and Planck's constant have dimensions, as does the magnitude of electrical charge ($e = 1.6 \times 10^{-19}$ coulomb). The numerical values of these quantities depend upon the units employed.

From these fundamental constants of nature one can form numerical quantities that are dimensionless or have dimensions of length. Do these have a deep significance?

One such is the ratio of proton and electron masses

$$m_\mathrm{p}/m_\mathrm{e} = 1836$$

which is a challenge to any theory of elementary particles. To date there is no explanation of this quantity.

Another interesting dimensionless quantity is

$$e^2/\hbar c \simeq 1/137$$

This involves electric charge e, $\hbar$ of quantum theory and the velocity of light c (which is important in relativity theory). Thus one might expect that this quantity will be important in a

Relativistic(c) quantum theory($\hbar$) of electric charge(e)

Such a theory exists—it is known as 'Quantum Electrodynamics'—and $e^2/\hbar c$ is the strength by which electrons couple to electromagnetic radiation. It is conventional to denote $e^2/\hbar c$ by the symbol α.

If m_p and m_e are proton and electron masses then the ratio of gravitational to electromagnetic forces in hydrogen atoms is

$$GMm/e^2 \simeq 10^{-40}$$

which quantifies our statement that gravity is exceedingly feeble in atomic and particle physics.

The gravitational constant, $\hbar$ and c form a quantity with the dimension of a length

$$\sqrt{(G\hbar/c^3)} \simeq 10^{-33}\,\mathrm{cm}$$

By analogy with quantum electrodynamics this suggests that the relativistic(c) quantum theory($\hbar$) of gravity(G) becomes important at distances of order $10^{-35}\,\mathrm{m}$ (or in energy equivalent, using the

Box 4.2—*continued*

uncertainty principle, at 10^{19} GeV). This is far beyond the reach of present technology and is even four orders of magnitude beyond the energies at which unified theories of the remaining natural forces are believed to be exact (Chapter 10). Therefore gravity may indeed be ignored in high energy particle physics (so far!).

Other important quantities with the dimensions of length include

$$\hbar/m_\pi c \simeq 10^{-13}\,\text{cm} \qquad (1)$$

and
$$\hbar/m_e c \alpha \simeq 10^{-8}\,\text{cm} \qquad (2)$$

The former (1) suggests that pions (p. 39) cannot be regarded as solid spheres at distances less than 10^{-13} cm—relativistic(c) and quantum effects(h) become important. This is manifested by the fact that pions transmit the nuclear force over this sort of distance. *This is the scale of size of light nuclei.* The latter length (2) involves the electron mass and the strength of the electromagnetic interaction. This is the distance scale at which electrons are typically held by electromagnetic attraction for the heavy nucleus. *This is the scale of size of hydrogen atoms.*

A collection of seemingly independent laws was found to govern various other electrical and magnetic phenomena.

Important progress took place in the middle of the 19th century with the discovery that electric and magnetic phenomena were intimately related. In 1820 Oersted discovered that a *magnetic* compass needle could be deflected when an *electric* current passed through a nearby wire; this was the first demonstration that electric currents could have magnetic effects. In 1831 Faraday discovered a complementary phenomenon: when a magnet was thrust into the centre of a coil an electric current spontaneously flowed in the coil. This showed that magnets could induce electrical effects. Faraday's discovery was eventually to lead to the development of electric generators and alternators, Bell's original telephone, transformers, and a variety of modern electrical techniques. It also gave crucial impetus to the study of the relation between electric and magnetic phenomena.

In 1864 Maxwell formulated his celebrated equations containing all of the separate laws responsible for these various phenomena. This united electricity and magnetism into what is now called 'electromagnetism' and in addition, predicted the existence of other phenomena which had previously been unsuspected, the most notable being electromagnetic radiation (see Box 2.2).

Maxwell's equations succinctly summarised all known electric and magnetic phenomena. In 1928 Dirac combined Maxwell's theory,

Box 4.3 Antimatter

In 1928 Paul Dirac produced his equation describing the motion of electrons in the presence of electromagnetic radiation. He found that he could only do this in a manner consistent with relativity if, in addition to electrons, there exist entities that appear to be positively charged electrons—these have identical properties to electrons apart from the opposite sign of electrical charge. This positive version of the electron is called a 'positron' and is an example of 'antimatter'.

Antimatter is a mirror image of matter, possessing equal and opposite sign of electrical charge but otherwise responding to natural forces much as the matter equivalent. Thus protons and antiprotons have the same mass and behave the same way, but have equal and opposite charges. Similarly, electrons and positrons have opposite charges but are otherwise essentially alike.

Matter and antimatter can mutually annihilate and convert into radiant energy. The amount of energy (E) is given by Einstein's famous equation

$$E = mc^2$$

where c is the velocity of light and m is the total mass annihilated. Conversely, if enough energy is contained in a small region of space-time then matter and antimatter will be produced in equal abundance. In the laboratory this is frequently done. High energy electrons and positrons annihilate to produce radiant energy which in turn produces new varieties of matter and antimatter. This is a popular way of creating and studying varieties of matter that are not abundant on Earth. Recently, discoveries of exotic matter with properties known as 'charm' were made in this way. At the start of the universe, in the heat of the Big Bang, there was such a concentration of energy that matter and antimatter were created from it in vast, equal, quantities. Yet today the universe is built of matter almost to the exclusion of antimatter: our atoms consist of electrons and protons, no examples exist of positrons orbiting antiprotons. Why this is so is an important puzzle.

Antimatter particles are conventionally denoted by a line over the symbol for the matter equivalent. Thus $\bar{p}$ is the antiproton and $\bar{\nu}$ the antineutrino. The positron is denoted e^- or more usually e^+. The positron was discovered experimentally by C. D. Anderson in 1932. The first example of antimatter was thus seen in the same year that the neutron was discovered!

relativity, and the newly-discovered quantum mechanics and showed that the resulting theory, 'quantum electrodynamics', enables one to calculate the effects that arise when light interacts with matter, in particular with electrically charged subatomic particles like the electron (see Box 4.3).

Box 4.4 **Feynman diagrams**

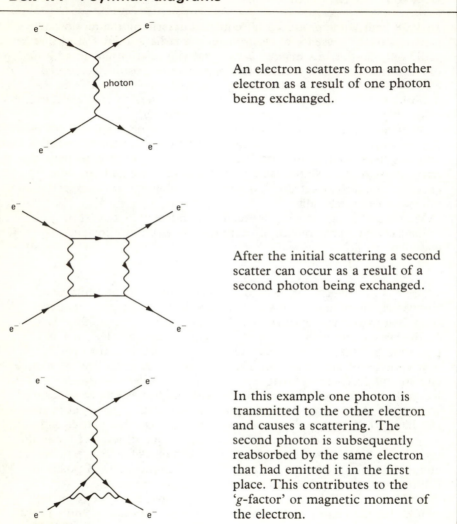

An electron scatters from another electron as a result of one photon being exchanged.

After the initial scattering a second scatter can occur as a result of a second photon being exchanged.

In this example one photon is transmitted to the other electron and causes a scattering. The second photon is subsequently reabsorbed by the same electron that had emitted it in the first place. This contributes to the '*g*-factor' or magnetic moment of the electron.

When an electrically charged particle accelerates, an electromagnetic wave is radiated. In quantum electrodynamics this wave behaves as if it were a series of particles, 'photons', and so we regard the particle's acceleration as resulting in the emission of one or more photons.

It is customary to draw a diagram to represent this (these are known as Feynman diagrams after their inventor Richard Feynman). An electron is represented by a straight line and a photon by a wiggly line. Time runs from left to right in the diagrams in Box 4.4 and so these represent an electron coming along and emitting one photon which is in turn absorbed by another electron. The photon transfers energy from the first to the second electron. On gaining this energy the electron is accelerated. Newton's laws of motion tell us that acceleration occurs when a force is applied, thus the photon has effectively transmitted a force whose origin was the original electron some distance away. In this sense we say that photons mediate electromagnetic forces.

This diagrammatic representation not only portrays in a conceptually helpful way what is happening, but can also be given a precise mathematical meaning. By associating specific mathematical expressions with the various lines and vertices that occur in these 'Feynman diagrams' high precision calculations can be made successfully for the electromagnetic properties of atomic particles.

Quantum electrodynamics has been tested time and again during the last 30 years. Its predictions have been verified to ten significant figures, a testament both to the ingenuity of experimentalists and to its validity as the theory of interactions between light and charged particles. It well deserves its acronym of QED.

Nuclear forces

Whereas the electromagnetic attraction of opposite charges is the source of the electrons' attraction for the nucleus, the mutual repulsion of like charges creates a paradox in the existence of the nucleus itself: many protons are packed tightly together despite experiencing the intense disruptive force of their mutual electromagnetic repulsions.

The attractive force that holds them there must be immensely powerful to overcome this electromagnetic repulsion. In deference to its strength it was named the 'strong' force. (Modern ideas imply that this is a misnomer, but I will adhere to it for historical reasons.)

One empirical feature of the strong force is that protons and neutrons experience it whereas electrons do not. This suggests that protons and neutrons possess some sort of 'strong charge' which electrons do not. By analogy, this led to the proposal that just as the electromagnetic force is carried by a photon so there should be a carrier of the strong nuclear force and the particle now called the pion was postulated.

From the observation that the nuclear force only acts over a few fermi (1 fermi = 10^{-13} cm) as against the potentially infinite range of electromagnetism, Yukawa (Box 4.5) computed that the pion had a mass about 1/7 of a proton mass (as against the masslessness of the photon). The eventual discovery of the pion in 1947 with this mass confirmed it as the carrier of the nuclear force over distances of the order of a fermi.

In nuclear decay involving α or γ emission, the strong and electro-

---

Box 4.5   Yukawa and the pion

In the large-scale world, energy appears to be absolutely conserved. Quantum mechanics shows that small amounts of energy (ΔE) can be 'borrowed' for a time (Δt) where

$$\Delta E \times \Delta t = \hbar = 6.6 \times 10^{-22} \, \text{MeV} \times \text{seconds}$$

This is at the root of the transmission of forces by the exchanging of particles (such as the photon in the case of electromagnetism). When a free electron emits a photon, energy is not conserved, the imbalance being determined by the amount of energy carried by the photon. The more energy the photon carries, so the sooner must that energy be repaid and the less distance the photon will have travelled before being absorbed by another charged particle and energy balance restored.

A photon has no mass so it is possible for the photon to carry no energy at all. In this case it could voyage for infinite time and so transmit the electromagnetic force over infinite distance. Contrast this with the nuclear force which binds nucleons to one another so long as they are less than about 10^{-12} cm apart, but does not act over larger distances.

This phenomenon led Yukawa to postulate that the carrier of the strong force had a mass. His reasoning was that energy and mass are related by Einstein's equation $E = mc^2$. Thus emission of the force-carrying particle would always violate energy conservation by at least mc^2 and hence the particle must be reabsorbed not later than time t:

$$t = \hbar/mc^2 \equiv \hbar/\text{some number of MeV}$$

Since it travels at less than the speed of light the maximum distance it can travel and transmit the force is:

$$\text{Max. distance} = ct = \hbar c/\text{mass in MeV}$$

Knowing that nuclear forces only occur over less than 10^{-12} cm led Yukawa to propose that their carrier (the 'pion') had a mass of about 140 MeV.

The subsequent discovery of the pion with this very mass led to Yukawa being awarded the Nobel Prize for physics in 1949.

---

magnetic forces are at work. In both of these types of decay, the number of neutrons and protons is separately conserved.

While these two forces control these and almost all other nuclear phenomena so far observed, there is one remaining process that they cannot describe. This is the source of the last of Becquerel's three radiations—the β-decay process.

Emission of beta particles (electrons) occurs when a neutron in the nucleus becomes a proton. The net electrical charge is preserved by the

Box 4.6A   Strength of the weak force I

Of the four known fundamental forces, electromagnetism and gravity have infinite ranges whereas the weak and strong nuclear forces operate only over very short distances, less than a fermi or so. As outlined in Box 4.5 the masslessness of the photon enables it to be emitted while at the same time conserving energy and momentum. The photon is thereby free to exist for ever and transmit electromagnetic forces over infinite distances. The production of particles with mass, such as the pion, for example, violates energy conservation and the larger the energy account is overdrawn so the sooner it must be repaid (uncertainty principle p. 16). Here is the source of the short range nature of the ensuing nuclear force between nucleons.

When a neutrino converts into an electron, a W^+ particle is emitted. As this is very massive the energy imbalance is huge and the length of time that this can be tolerated is, in turn, very short indeed. The distance over which the W can travel and transmit the weak force is therefore very short. Only extremely rarely does it voyage even 1 fermi and so it is very unlikely that particles will feel this force as against, say, electromagnetism whose effects are easily transmitted over such a distance.

The feeble strength of the 'weak' force is believed to result directly from the huge mass of its carriers—the W^+, W^- particles—which obscures the possibility that the intrinsic strength of the W coupling to the electron is comparable to electromagnetism's. Once this is realised, the weak and electromagnetic forces no longer appear to be very different from one another. This is at the root of the modern ideas of uniting them into a single theory of the 'electroweak' force—combining electromagnetism and the weak force within it.

W bosons and the weak force

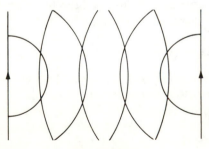

Low energies
Chance of radiating electromagnetic (γ) or weak (W) the same.

Electromagnetic wave
(photon)

Box 4.6A—*continued*

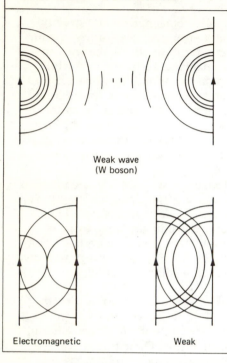

Weak wave
(W boson)

Electromagnetic   Weak

The chance of γ propagating and transmitting the (electromagnetic) force between distant particles is much greater than a W particle's chance of transmitting the weak force. The range of the weak force is of order $\hbar/mc$ where $\hbar$ is Planck's constant and c the velocity of light. When $m = m_w \simeq 100$ GeV, then the range is only of the order of 10^{-16} cm.

High energies
At very high energies the particles can approach one another very closely.

The chances of γ or W crossing this gap are now quite similar. The strength of the 'weak' force is now comparable to electromagnetism.

Box 4.6B   Strength of the weak force II

The strength of the electromagnetic force is expressed by the dimensionless quantity $a \simeq 1/137$. The strength of the weak force is expressed in terms of G_F where:

$$G_F \simeq 10^{-5}/m_p^2 \qquad (m_p = \text{proton mass} \simeq 1 \text{ GeV})$$

known as the Fermi constant after Enrico Fermi who made the first attempt at constructing a theory of the weak force (Chapter 8). Note that G_F has dimensions with the result that

$$G_F/a \simeq 10^{-3}/\text{GeV}^2$$

Thus in nucleon beta decay and manifestations of this force at 1 GeV energy, the relative strength as compared to electromagnetic forces is

$$G_F/a \approx 10^{-3}$$

hence the naming of it as the 'weak' force.

But this is misleading. It is weak in its effects at 1 GeV, which is where experiments were restricted for a quarter of a century after 1935. However, today we can study the force with high energy neutrinos and

---

Box 4.6B—*continued*

---

the *dimensionless* measure:

$$\frac{G_F}{\alpha} \times \frac{(\text{Energy of experiment})^2}{m_p^2}$$

can be of order unity. More precisely the comparison that is relevant is:

$$\frac{G_F}{\alpha} \frac{m_w^2}{m_p^2}$$

where m_w is related to the mass of the weak force carrier (the analogue of the photon in electromagnetic forces). If m_w is some tens of GeV then the strengths of 'weak' and electromagnetic forces are comparable.

Modern ideas imply that the weak and electromagnetic forces are two manifestations of a single 'electroweak' force and that the carrier of the weak force has mass about 80–90 GeV. This carrier was discovered in 1983 at a new high energy accelerator at CERN Geneva. More details can be found in Chapter 8.

---

emission of the electron

$$n^0 \rightarrow p^+ + e^- + \overline{\nu^0}$$

but the number of neutrons, protons, and electrons is not. The electromagnetic and strong forces do not have such an ability: the number of neutrons and protons or electrons is conserved when these forces act. The agent responsible for the neutron's β-decay is known as the 'weak' nuclear force, being some hundred thousand times less powerful than the strong nuclear force when acting in this way (see Box 4.6A and B).

An important property of β-decay is the fact that it produces neutrinos (ν^0) (technically antineutrinos $\overline{\nu^0}$, Box 4.3). These have no charge and so do not feel electromagnetic forces and, like the electron, do not experience the strong nuclear force. The only force that measurably affects them is the weak force. This blindness to the other forces makes neutrinos a unique tool for studying the weak force; by firing them at targets and studying how their flight is disturbed we are seeing directly the weak force at work (see Box 4.7).

When neutrinos are fired at matter they are most noticeably converted into electrons. Recently, examples of the action of the weak force have been seen where neutrinos are scattered without changing their identity. There are similarities between these processes and the familiar electromagnetic scattering of electrons from charged matter. The modern view is that the electromagnetic and weak forces are probably two manifestations of a single 'electroweak' force.

Box 4.7   Parity:mirror symmetry

If mirror symmetry was an exact property of nature then it would be impossible to tell whether a film of an experiment has been made directly or by filming the view in a mirror into which the experiment had been reflected. This is equivalent to saying that nature does not distinguish between left and right in an absolute way. This is indeed the case for phenomena controlled by gravitational, strong, and electromagnetic interactions. As these control most observed phenomena it had been assumed that left–right symmetry was an inherent property of all subatomic processes. But in 1956 mirror symmetry was discovered to be broken in weak interactions.

The historic experiment involved the beta decay of cobalt nuclei but we can illustrate it for the beta decay of a single neutron. An electric coil gives rise to a magnetic field which interacts with the neutron's magnetic moment and aligns its spin along the direction of the field's axis. The electrons produced in the decay are preferentially emitted upwards (a). Viewed in a mirror, the electrons are emitted upwards if an electric current flows in the coil in the *opposite* direction (b). If mirror symmetry is a property of nature, the electrons should still be emitted upwards in the lab when the current flows in the opposite direction. However, what is observed is that the electrons are emitted downwards (c). More precisely, the electrons are emitted on that side of the coil from which the current is seen to flow clockwise. By this violation of mirror symmetry, nature provides an absolute meaning to left and right. If we imagined a magic mirror that also interchanged matter and antimatter then the combined exchange of left–right and matter–antimatter would restore symmetry in this process (d).

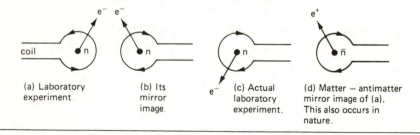

(a) Laboratory experiment.

(b) Its mirror image.

(c) Actual laboratory experiment.

(d) Matter – antimatter mirror image of (a). This also occurs in nature.

Just as the photon is the carrier of the electromagnetic force and the pion the carrier of the strong nuclear force, so there should be a carrier of the weak force. When neutrinos are converted into electrons, this is believed to be due to the action of an electrically charged force carrier, the 'W boson'. When neutrinos scatter and preserve their identity it is a neutral 'Z boson' believed to be responsible. They were predicted to be nearly 100 times more massive than protons—too large for them to be produced in accelerators, until now. However, a new machine at CERN Geneva has just come into operation at which we hope to produce W and Z particles.

Early in 1983 the first sighting of a W boson was reported—weighing in at over 80 proton masses. At the time of writing the Z boson is still being sought.

The history of the weak interaction theory and more about the W and Z bosons will be deferred until Chapter 8, after the search for the strong force's carrier, the pion, has been described. The discovery of the pion will open up a Pandora's box that will confuse physicists for nearly two decades. When the dust settles our story will have reached the 1960s. A deeper layer of matter consisting of 'quarks' will be perceived inside nuclear matter. Not until then will we be able to contemplate a complete theory of the weak force.

5 *Nuclear particles and the Eightfold Way*

Pions

In 1935 Yukawa proposed the existence of a particle as the carrier of the force that gripped neutrons and protons in the nucleus (see Box 4.5). This is called a pi-meson, or pion, labelled π. The resulting exchange of the pion between two protons or two neutrons can be portrayed as in Box 5.1.

The neutron has emitted a pion and remained a neutron in (a) and the proton emitted a pion and stayed a proton in (b). In each example the pion has no electrical charge (the total charge is always conserved; the neutron and proton gave none up, so the pion carries none). To denote its electrical neutrality we label it π^0.

A pion can be emitted by a neutron and absorbed by a proton, or vice versa (c) and so cause the neutron and proton to exert a force on one another. Yukawa's theory of the nuclear force also required the possibility that the neutron and proton exchanged their positions (d) or (e). There is still a neutron and a proton at the start and at the finish as in (c), but this time electrical charge has been carried across. If the forces between neutrons and protons are transmitted by pions then *three* varieties of pion are called for. These are denoted π^+, π^-, π^0 depending upon whether they carry positive, negative, or no electrical charge respectively, and their masses are predicted to be almost identical to one another. In contrast to protons which are stable in the nucleus*, pions survive for less than 10^{-8} seconds before decaying into photons, neutrinos and electrons or positrons. (See Box 4.3.)

The best hope of finding pions seemed to lie in the cosmic ray showers that continuously bombard the Earth. Cosmic rays consist of heavy nuclei, protons, electrons, and similar objects produced in stars and accelerated to extreme energies by magnetic fields in space. By studying them in detail it was hoped to discover other types of matter that had not previously been observed on Earth.

---

* It is possible that protons might not be permanently stable (see Chapter 10).

Box 5.1 Exchange of a pion between neutrons and protons

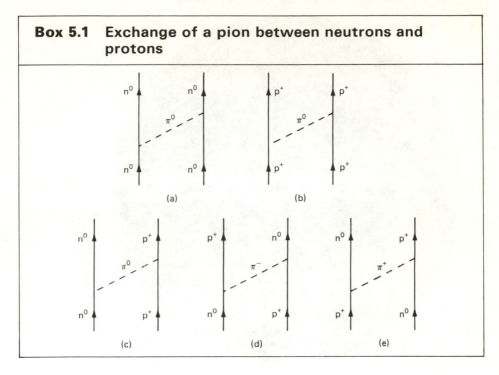

Originally, cosmic rays were studied using a cloud chamber, a device devised by C. T. R. Wilson in 1911. When an electrically charged particle (such as a cosmic ray) passes through supersaturated vapour, it ionises the vapour's atoms. Droplets of water settle on these ions forming a vapour trail similar to that from a high-flying aircraft. The trajectory of the particle is thereby revealed.

From the form of the trail it is possible to learn about the nature of the particle that has passed through. Massive particles (such as nuclei) plough straight through and cause many drops to form, yielding a thick, straight track. Electrons are so light that collisions with the atoms in the chamber easily divert them from their path and a rather diffuse, wavy track results.

If a cloud chamber is placed in a strong magnetic field, the trajectory of charged particles will be bent. Light particles are deflected more easily than heavy ones and the direction of the deflection shows whether the charge on the particle is positive or negative. Thus a lot of information about charged particles can be easily obtained. Uncharged particles, on the other hand, leave no track and their presence can only be detected indirectly (compare the invisible man, p. 22).

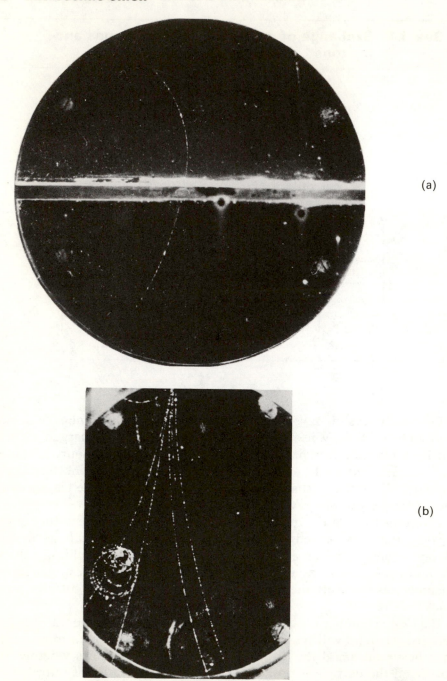

(a)

(b)

Photo 5A *Antimatter in a cloud chamber* (*1932*)   (a) This cloud chamber track gave the first clear evidence for the existence of a particle with the mass of the electron but with *positive* electrical charge—the 'positron' or 'antimatter electron'. The cloud chamber is 16 cm in diameter and 4 cm deep; the bright

Box 5.2   The bubble chamber

Donald Glaser was gazing at a glass of beer, watching the bubbles rise to the surface and musing about the minute imperfections of the glass container on which the bubbles form. From this contemplation the idea of the bubble chamber was born.

In a bubble chamber the paths of charged particles are made visible. The chamber contains liquid that is on the point of boiling. A piston in the chamber is used to lower suddenly the pressure of the liquid causing the liquid to start boiling. Bubbles of gas start to form but to grow they need some central stimulus such as the irregularities on the surface of the beer glass. If a charged particle passes through the liquid at the critical moment it ionises some of the liquid's atoms and these act as centres for bubble formation. The trail of bubbles reveals its trajectory.

Sometimes the particle interacts with one of the protons or neutrons in the liquid and produces new particles. These show up as multiple tracks emanating from the point of interaction. In this way production and properties of the particles are studied and, occasionally, new varieties of particles are produced and detected. The earliest bubble chambers were of the order of a metre in diameter but today they may be as large as that shown in Photo 5B!

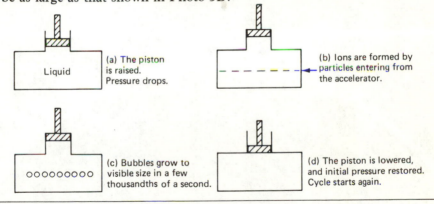

(a) The piston is raised. Pressure drops.

Liquid

(b) Ions are formed by particles entering from the accelerator.

(c) Bubbles grow to visible size in a few thousandths of a second.

(d) The piston is lowered, and initial pressure restored. Cycle starts again.

line across the diameter is a lead plate. A magnetic field of 15 000 gauss causes charged particles' tracks to be curved. In this field, negatively charged particles follow clockwise trajectories, and positively charged follow anticlockwise ones, hence this is either an upward-moving positive charge or downward-moving negative charge. As the particle slows so its trajectory becomes more curved. The greater curvature at the top shows that this particle entered from below and has an anticlockwise trajectory—positive charge.

Any doubt is removed in photo (b) where energy, probably carried by photons, has converted into equal amounts of matter and antimatter just outside the top of the chamber. Luckily the products all passed through—three electrons follow clockwise trajectories and three antielectrons follow anticlockwise paths (the middle track of these three is not well-illuminated and escapes from view halfway down the chamber). (*Photos Science Museum, London by courtesy of Professor C. D. Anderson*).

(a)

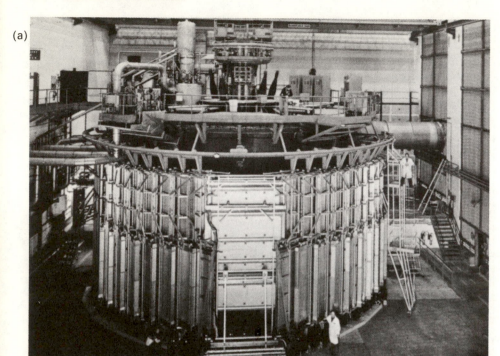

(b)

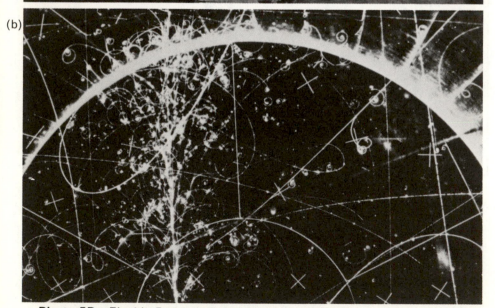

Photo 5B *The big European bubble chamber* (*BEBC*)

(a) BEBC at CERN (note the size of the people) can reveal complex collisions between subatomic particles and render them visible as seen in part (b).

(b) This photo is looking into the cylinder of BEBC from above and reveals a complicated shower of subatomic particles. (*Photo CERN*)

Cloud chambers have been superceded by bubble chambers. Instead of trails of water drops in a cloud, the tracks in a bubble chamber consists of bubbles formed in a superheated liquid (see Box 5.2). The density and changes in direction of the tracks in a bubble chamber yield information on the particles' properties analogous to that in a cloud chamber. However, the bubble chamber has several advantages, in particular that it is both target and detector. Collisions between the projectile particles and the nuclei in the bubble chamber liquid can be directly observed, and not only the original particle but also the nuclear fragments can be seen (Photo 5B). The art of the experimentalist is to study the resulting tracks and determine what particles caused them.

What will one of the pions look like? A charged pion is heavier than an electron but lighter than a proton. Hence on passing through a cloud chamber or bubble chamber in the vicinity of a magnet it will curve less than the light electron but more than the massive proton. The direction of the bending will be the same as that of an electron in the case of π^- but the same as that of a proton in the case of a π^+ (the π^- and π^+ bending in opposite directions).

In 1936, C. D. Anderson and S. H. Neddermeyer found such a track in a cloud chamber. The particle responsible had a positive charge, and a mass that was slightly lighter than Yukawa's prediction. A negative version of the particle was found by J. Street and E. Stevenson at about the same time. However, no evidence for an electrically neutral partner turned up. This was a puzzle. More disturbing was the fact that the particle showed no desire to interact with nuclei. As the *raison d'etre* for Yukawa's particle had been that it provided the grip that held the nucleus together, then it must necessarily have a strong affinity for nuclear material.

The resolution of the puzzle was that this particle was *not* Yukawa's particle, nor did it have any role as a carrier of the strong nuclear force.

We now know that what Anderson discovered in 1936 was a muon (denoted μ)—a particle which is similar in many ways to an electron except that it is about 200 times heavier. This was quite unexpected and the reasons for its existence remained a total mystery for the next 40 years.

Yukawa's particle, the pion, was finally discovered in 1947 by C. F. Powell, who suspected that interactions in the atmosphere might be preventing most pions from ever reaching the Earth. So he placed stacks of photographic emulsion on the Pic du Midi in the French Pyrenees. Cosmic rays passing through affected the chemical and produced a dark track when the photographic emulsion was developed. The mass of the pion was found to be 140 MeV, see Box 3.8, as Yukawa had predicted.

Box 5.3   More than one variety of neutrino

When a π^+ decays it usually produces a positively charged particle called a muon ('antimuon') and a neutrino. Less often it will produce an antielectron (positron) and a neutrino. These two neutrinos are not identical, we denote them ν_μ and ν_e to show that they were born in conjunction with μ or e respectively. If these neutrinos hit neutrons then they can convert a neutron to a proton plus an electron or muon:

$$\nu_e + n \rightarrow p + e^-$$

$$\nu_\mu + n \rightarrow p + \mu^-$$

The neutrino that was born with an electron *always* produces electrons, never muons. Similarly the ν_μ *always* produces muons, never electrons. In all known weak interaction processes, the ν_e and electron are paired and the ν_μ is paired with μ. Recently a third variety of charged particle has been found, with properties similar to the e^- and μ^-. This is known as the tau, τ^-. It appears to be paired with a third variety of neutrino (ν_τ) in weak interactions.

How it is that these neutrinos 'know' of their separate identities is not yet understood. We do know that they are distinct and appear to be siblings of the electron, muon, and tau. This pairing appears to be a fundamental part of nature's pattern.

Box 5.4   Particle summary 1947

Name/Symbol	Charge	Mass (proton as unit)	Stable	Does it feel the strong force?
Electron (1897) e	-1	$\frac{1}{1800}$	yes	no
Neutrino (1931–56) ν	0	0 (?)	yes	no
Muon (1936) μ	-1	$\frac{1}{9}$	$\tau \sim 10^{-6}$ s★	no
Proton p	$+1$	1	yes (?)	yes
Neutron (1932) n	0	1	$\tau \sim 15$ min when free	yes
Pion (1947) $\pi^\pm$	± 1	$\frac{1}{7}$	$\tau \sim 10^{-8}$ s	yes
[(1950) π^0	0	$\frac{1}{7}$	$\tau \sim 10^{-16}$ s	yes]

★ The symbol τ means lifetime.

The π^+, π^-, and π^0 had been predicted by Yukawa. Following the $\pi^\pm$ discovery in 1947 the observation of the uncharged sibling, π^0, was confidently awaited. I include it in this list for completeness although it was not directly detected until 1950. Similar comments apply to the neutrino—accepted by the physics community but not detected until 1956 (Chapter 3). Two varieties of antimatter had also been detected: the antielectron (positron e^+ discovered in 1932 and the antimuon, μ^+, discovered in 1936.

When the π^+ or π^- decay they usually produce a μ^+ or μ^- and a neutrino. See Box 5.3. Ironically the muon detected by Anderson in 1936 was the progeny of the particle that he had been seeking.

The π^+ and π^- were finally produced in particle accelerators at Berkeley in 1948 as products from the collision of alpha particles and carbon nuclei. The π^0 was subsequently found in 1950 as a product of similar collisions.

So apart from the unexpected appearance of the muon everything was turning out rather well. Box 5.4 summarises the situation in 1947.

Strange particles

As studies of cosmic rays continued, so further particles were discovered. These were produced by the strong interaction of pions with protons and neutrons in the atmosphere. Having been readily produced this way, one would have expected that these new particles would have rapidly decayed back into pions and protons, the very particles which had been responsible for their production. However, this did not occur. Another peculiar property was that they were always produced in pairs. These unusual properties in production and decay caused them to be dubbed 'strange' particles.

Among the strange particles are mesons somewhat heavier than pions. These are the electrically charged 'kaons', K^+ and K^-, and two varieties of neutral kaon denoted K^0 and $\overline{K}^0$ all with masses of about 500 MeV. The discovery of the K^0 is usually attributed to Rochester and Butler who found it in cosmic rays in 1947 (Photo 5C). The uncharged K^0 left no track of its own but decayed into two charged particles (now known to be π^+ and π^-) and so left a distinctive V-shaped vertex as in the

---

Box 5.5   Hadrons and leptons

As the number of particles proliferated, attempts were made to organise them into families with common properties. Some particles (such as the electron and neutrino) do not experience the strong interaction. These are called **leptons**. (The name is taken from that of a small Greek coin.) Nuclear particles feel the strong interaction and are known as **hadrons**. Hadrons are subdivided into two categories—*mesons* (such as the pion) and *baryons* (such as proton).

Hadrons possess an intrinsic angular momentum known as 'spin' which is a multiple of Planck's constant $\hbar$ in magnitude. For baryons this multiple is a half-integer: $\frac{1}{2}, \frac{3}{2}, \frac{5}{2} \ldots$, whereas for mesons it is an integer: $0, 1, 2 \ldots$. All leptons so far discovered have spin $\frac{1}{2}\hbar$.

(a)

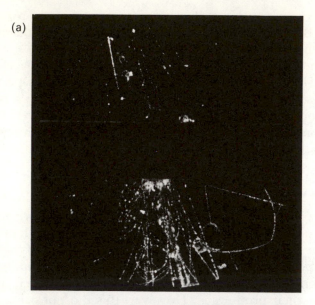

(b)

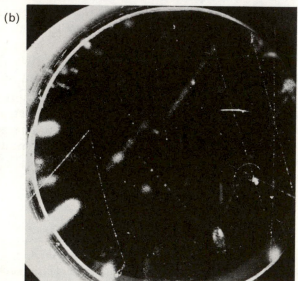

Photo 5C *V particles in a cloud chamber (1947)*

(a) The first photograph of a 'V particle' decay is of a K°. The resulting V is to the right of the picture immediately below a 5 cm lead plate traversing the chamber. The bulk of the tracks are pions and protons, in a shower of cosmic ray particles. The angle between the particles in the V is much too large for the V to be electron and positron (contrast part (b)) and is most probably $\pi^+\pi^-$.

(b) Here we see the decay of a $\Lambda^\circ \rightarrow p^+\pi^-$ on the left-hand side of the chamber. The massive proton gives the dense track with little curvature, the lighter pion yields a curving, less dense trajectory. (*Photo (a) Science Museum, London, courtesy Professor C. C. Butler; photo (b) Professor Butler.*)

illustration. This caused them initially to be referred to as V particles. It subsequently turned out that K mesons had been seen, but not recognised, in cloud chamber photographs of cosmic rays studied by Laprince-Ringuet and Lheritier as early as 1944.

In any reaction initiated by pions or nucleons the strange kaons are always produced in partnership with other strange particles such as the lamda (Λ), sigma (Σ) or xi (Ξ). These particles are more massive than the proton and neutron and from their properties we now recognise them as strange baryons (Box 5.5), with baryon number $+1$ the same as the proton. The striking feature of the pair production can be illustrated as follows.

If a negatively charged pion hits a proton, then baryon number (B) and charge (Q) conservation (Box 5.6) would allow both

$$\pi^- + p \rightarrow K^- + \Sigma^+ \qquad (B_{\text{total}} = 1, Q_{\text{total}} = 0)$$

and

$$\pi^- + p \rightarrow \pi^- + \Sigma^+ \qquad (B_{\text{total}} = 1, Q_{\text{total}} = 0)$$

to occur. Indeed it is energetically easier to produce light pions than the heavier kaons, so the second reaction should be more copious than the first. The puzzle was that this reaction has never been seen whereas millions of examples of the former have been observed.

To rationalize this Gell-Mann, Nakano, and Nishijima in 1953 proposed that there exists a new property of matter which they named 'strangeness' and that this strangeness is conserved in strong interactions of hadrons (Box 5.5). A pion or proton has no strangeness. When they interact and produce a particle with strangeness $+1$ then another particle with strangeness -1 must be produced in association so that the total strangeness is conserved. Thus if we arbitrarily assign strangeness value $+1$ to the K^+ we can deduce the strangeness of all other particles from a study of which reactions are, or are not, observed. Thus the Σ^- for example has strangeness -1:

$$\pi^0 + n \rightarrow K^+ + \Sigma^-$$

Strangeness: $\qquad 0 + 0 \qquad +1 + -1 \qquad$ (net strangeness zero)

and the unobserved reaction is forbidden because strangeness would not be conserved.

$$\pi^- + p \nrightarrow \pi^- + \Sigma^+$$

Strangeness: $\qquad 0 + 0 \qquad 0 - 1$

This scheme can be applied to all strange particles and a totally consistent picture emerges. The Λ and Σ have strangeness -1, the Ξ has strangeness -2, the K^+ and K^0 have strangeness $+1$ while the K^- and $\overline{K^0}$ have strangeness -1.

Box 5.6   Charge and baryon number conservation in particle interactions

---

Charge

Charge is expressed in units of the proton's charge. All particles so far detected have charges that are integer multiples of this quantity. The total charge never changes in any reaction or decay, for example in

$$\pi^- + p \to \pi^0 + n^0$$

charge (Q): $-1 + +1 \to 0 + 0$

the totality is preserved at zero. Thus charge conservation forbids

$$\pi^- + p \to \pi^+ + p$$

which has indeed never been seen.

Baryon number

Electrical charge conservation does not prevent

$$p^+ \to e^+ + \pi^0$$

but this decay of the proton has not been seen: the proton is very stable with a half-life that is at least twenty orders of magnitude greater than the life of the present universe! It is possible that protons are absolutely stable. This has been rationalised by inventing the principle of baryon conservation.

The baryon number of the proton and neutron is defined as $B = 1$; that of the lepton or photon is defined to be zero. Mesons, such as π, also have zero baryon number. Thus

$$p \to e^+ + \pi^0$$
$$B: 1 \quad\quad 0 \quad\quad 0$$

is forbidden because the baryon number changes. (Recently there have been theoretical suggestions that baryon number might not be conserved after all, but there is no evidence to support this yet.)

Strange particles also are assigned baryon number. Particles such as Σ, Λ, Ξ which decay and ultimately leave a proton in their decay products have $B = 1$. The K and other strange mesons decay into pions, photons, and leptons: they are assigned $B = 0$. Baryon conservation applies to them as well as to nonstrange particles.

At this point you may well ask 'But what is strangeness?' It is a property of matter, analogous to electric charge, which some particles have and others do not. This may sound rather an arid answer but it is all that physics can say at present. Physicists invent concepts and rules to enable them to predict the outcome of natural processes. By inventing strangeness we can successfully predict which reactions will or will not

Box 5.7 The strange life

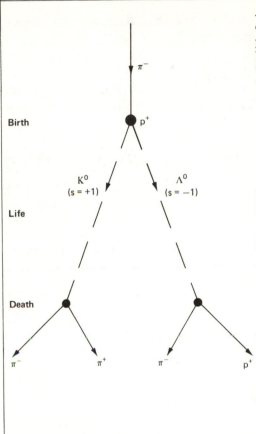

Birth

Life

Death

K^0
$(s = +1)$

Λ^0
$(s = -1)$

π^-

p^+

π^- π^+ π^- p^+

A pion incident from a cosmic ray or particle accelerator hits a proton in the atmosphere or a bubble chamber.

Two particles are produced. The reaction is written:

$$\pi^- + p \to K^0 + \Lambda^0$$

Strangeness

$$0 + 0 \to (+1) + (-1)$$

Strangeness is balanced by production of a pair of strange particles.

Strangeness likes to be conserved. This prevents the K^0 or Λ^0 dying.

Eventually weak interaction steps in. Strangeness is no longer conserved.

Electrical charge is always conserved. The neutral K^0 and Λ^0 each produce a pair of charged particles. These show up in photographs as a characteristic V.

The Λ baryon produces a proton in its decay; the K meson decays to π mesons: baryon number is conserved.

occur. Although the question of what strangeness 'is' is currently metaphysics, insights have been gained into the reason why the various particles carry the particular magnitudes of strangeness that they do (Chapter 6).

One further property of strangeness needs mention and that concerns its role in hadron decay. The Σ^0 baryon is more massive than the Λ^0 and so can lose energy by radiating a photon (γ) and converting into a Λ^0 while conserving strangeness

$$\Sigma^0 \to \Lambda^0 + \gamma$$

which the Σ^0 does within 10^{-20} seconds of its birth. But the Λ^0 is the lightest strange baryon, it cannot decay into lighter particles if baryon number and strangeness are both to be conserved. Thus we would expect the Λ^0 to be absolutely stable.

In fact it is metastable. After about 10^{-10} seconds it decays and violates strangeness. One of its decay modes is

$$\Lambda^0 \rightarrow p^+ + e^- + \overline{\nu^0}$$

the initial particle having strangeness -1, the final state having none. Similarly the K^-, for example, decays and violates strangeness as follows.

$$K^- \rightarrow \mu^- + \overline{\nu^0}$$

These decays have similar behaviour to those of the neutron and the lightest mesons

$$n^0 \rightarrow p^+ + e^- + \overline{\nu^0}$$
$$\pi^- \rightarrow \mu^- + \overline{\nu^0}$$

which are well known manifestations of *weak interactions*. Thus it appears that strangeness is conserved in strong interactions and violated in weak interactions (Box 5.7 and 5.8A).

Box 5.8A   Known and postulated hadrons 1953

Name/Symbol	Charge	Mass (proton as unit)	Strangeness	Stable
Proton p	$+1$	1	0	Yes (?)
Neutron n	0	1	0	$\tau \sim 15$ min free
Pion $\pi^{\pm}$	± 1	$\frac{1}{7}$	0	10^{-8} s
π^0	0	$\frac{1}{7}$	0	10^{-16} s
Kaon $K^{\pm}$	± 1	$\frac{1}{2}$	± 1	10^{-8} s
$K^0, \overline{K}{}^0$	0	$\frac{1}{2}$	± 1	10^{-8} or 10^{-10} s
Sigma $\Sigma^{\pm}$	± 1	1.2	-1	10^{-10} s
[(1958) Σ^0	0	1.2	-1	10^{-20} s]
Lambda Λ^0	0	1.1	-1	10^{-10} s
Xi [(1959) Ξ^0	0	1.3	-2	10^{-10} s]
Ξ^-	-1	1.3	-2	10^{-10} s

The Λ^0 was too light to be mistaken for the uncharged sibling of the Σ^+ and Σ^-: a Σ^0 was predicted. The Gell-Mann and Nishijima strangeness scheme required that an uncharged Ξ partnered the observed Ξ^-. The Σ^0 and Ξ^0 were observed in 1958 and 1959 respectively. There was much confusion in understanding the $K^0, \overline{K}{}^0$ mesons which was not resolved until 1956. Baryons are shown in ordinary and mesons in bold type.

Box 5.8B   Hadron summary *c.* 1960

Name/Symbol	Charge	Mass (proton as unit)	Strangeness	Stable
Proton p	$+1$	1	0	Yes (?)
Neutron n	0	1	0	$\tau \sim$ 15 min free
Pion $\pi^{\pm}$	± 1	$\frac{1}{7}$	0	10^{-8} s
π^0	0	$\frac{1}{7}$	0	10^{-16} s
(1961) **Eta** η^0	0	$\frac{1}{2}$	0	10^{-19} s
Kaon $K^{\pm}$	± 1	$\frac{1}{2}$	± 1	10^{-8} s
$K^0, \overline{K^0}$	0	$\frac{1}{2}$	± 1	10^{-8} or 10^{-10} s
Sigma (1958) $\Sigma^{\pm}$	± 1	1.2	-1	10^{-10} s
Σ^0	0	1.2	-1	10^{-20} s
Lambda Λ^0	0	1.1	-1	10^{-10} s
Xi (1959) Ξ^0	0	1.3	-2	10^{-10} s
Ξ^-	-1	1.3	-2	10^{-10} s
Delta Δ^{++}	$+2$	1.2	0	10^{-23} s
Δ^+	$+1$	1.2	0	10^{-23} s
Δ^0	0	1.2	0	10^{-23} s
Δ^-	-1	1.2	0	10^{-23} s
(1961) Sigma-star				
$\Sigma^{\pm}\star$	± 1	1.4	-1	10^{-23} s
$\Sigma^{0}\star$	0	1.4	-1	10^{-23} s
(1962) Xi-star				
$\Xi^{0}\star$	0	1.5	-2	10^{-23} s
$\Xi^{-}\star$	-1	1.5	-2	10^{-23} s

The discoveries of Σ^0 and Ξ^0 completed an octet of baryons. The eta meson discovery in 1961 showed that mesons formed a family of eight analogous to the eight baryons. The ingredients of Yuval Ne'eman and Murray Gell-Mann's 'Eightfold Way' theory (1961) was then to hand. This theory predicted that a family of ten baryons should exist. The $\Sigma\star$ and $\Xi\star$, announced in 1962, led to Gell-Mann's dramatic prediction of the Ω^- particle (Box 5.12) and the verification of the theory. Mesons are in bold type.

More hadrons

During the 1950s and 1960s the advent of high energy accelerators (Box 5.9) enabled high velocity protons to be fired at nuclear targets. The debris that emerged from these collisions contained new varieties of matter that survived for only 10^{-23} seconds, the time that it takes for light to cross a proton. For example there were the Δ baryons Δ^-, Δ^0, Δ^+, Δ^{++} which produce protons in their decays, such as $\Delta^+ \to p\pi^0$. These delta particles have no strangeness. Their masses are about 1235 MeV, 30% greater than the proton.

Box 5.9 Modern particle accelerators

Early examples of particle accelerators were described in Box 3.2. By 1945 a large machine had been built which was dubbed a 'synchrocyclotron'. It accelerated protons up to 720 MeV energy. Smashing these into nuclear targets led, among other things, to the production of the π^0 meson in the laboratory. The solid magnet of that machine has been superceded by magnets surrounding an evacuated tube along which the protons can be accelerated.

In 1953 the first machines were built capable of accelerating protons to over 1 GeV. These devices at Brookhaven, New York, and Berkeley, California, enabled the massive strange particles to be produced in the laboratory—previously one had had to study the uncontrollable cosmic rays. The advent of these machines led to the discovery of baryon and meson octets.

Synchrocyclotrons which accelerated protons up to 30 GeV were built at CERN, Geneva and at Brookhaven in the mid-1950s and a whole spectrum of hadrons, with masses up to 3 times greater than the proton, emerged. At the Fermi laboratory near Chicago and at CERN one can now accelerate protons up to 500 GeV in a ring whose diameter is over a mile. The CERN machine is known as the super-proton-synchrotron or SPS.

Photo 5D *The British 'Nimrod' accelerator (1960s)* The proton synchrocyclotron *Nimrod* at Britain's Rutherford Laboratory. This could accelerate protons to an energy of 7 GeV, and played a major role during the 1960s in establishing the existence of Eightfold Way patterns. (*Rutherford Laboratory Photo*).

Box 5.9—*continued*

(a)

(b)

Photo 5E *CERN*

(a) The super proton synchrocyclotron at CERN Geneva (1978). This can accelerate protons up to 500 GeV. The machine is a circle of magnets so large that the bending is barely visible round the corner. (*Photograph CERN*)

(b) CERN is a testament to successful international collaboration. The enormous effort involved in probing nature at minute distances is seen in the size of the CERN site. To the right of the water tower are the CERN Intersection Storage Rings where protons collide head-on. The synchrocyclotron is in the middle of the picture. The 500 GeV SPS is off to the right of the site (not shown). (*Photo CERN*).

$\Sigma^\star$ particles were observed with similar rapid decays into Σ and π. The properties of the decays showed them to be very similar to that of Δ into p and π. Strangeness conservation in the rapid decays shows that $\Sigma^\star$ particles have strangeness minus one. Their masses are about 1385 MeV.

Strangeness minus two particles were found, $\Xi^{\star-}$ and $\Xi^{\star 0}$, which decayed rapidly into Ξ^- and Ξ^0. Analogous arguments to the above require these to have the same strangeness as the Ξ, namely minus two. The $\Xi^\star$ masses are about 1530 MeV.

The apparent simplicity and order that had existed in 1935, when it was thought that only a handful of elementary particles existed, had been replaced by a new complexity, see Box 5.8B. But then a pattern of regularity was noticed in the properties of this rapidly growing 'zoo' of particles and a new simplification emerged. What Mendeleev had done in 1869 for the atomic elements so Murray Gell-Mann and Yuval Ne'eman did for the subnuclear particles nearly a century later, in 1960–61.

The Eightfold Way: 'A periodic table for the nuclear particles'

To illustrate what Gell-Mann and Ne'eman did, we will begin with the mesons (pions and K particles). Mesons have electrical charges of zero, $+1$, or -1. They also have strangeness of zero (the pions) or $+1$ (the K^+ and K^0) or -1 (the K^- or $\overline{K^0}$). We could draw a diagram with the amount of strangeness on the vertical axis and the amount of charge along the horizontal (for historical reasons the charge axis is at an angle—Box 5.10). We now place the mesons at various points on this figure.

The K^+ has strangeness plus one and charge plus one. This is the point at the top right-hand corner (where the line for charge plus one intersects the horizontal line for strangeness plus one). So we place the K^+ at this point of the figure. The place that the line for charge plus one intersects the line for strangeness zero is at the far right of the figure. The particle with no strangeness and with charge plus one is the π^+ and so we put the π^+ there.

Continuing in this way we find a position for each and every particle and the resulting pattern is shown in Box 5.10 part (c). The pattern is a hexagon with a particle (π^0) at the centre.

Now we can play the same game for the baryons (the neutron, proton, Λ, Σ, and Ξ) but add one unit to the strangeness axis. The same hexagonal structure emerges when we place the particles on the figure but this time we find two particles at the centre instead of the one in the previous example (Box 5.11).

Box 5.10 The Eightfold Way

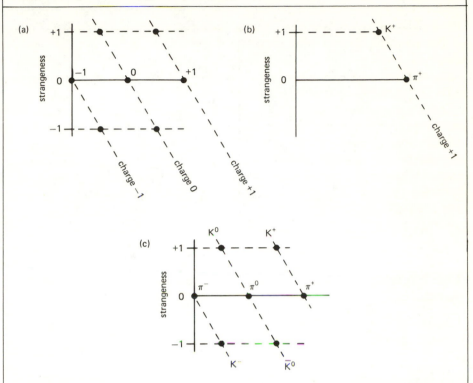

Hadrons—particles that interact through the strong nuclear force—fall into patterns according to properties such as electric charge (plotted horizontally) and strangeness (plotted vertically). These patterns are called the 'Eightfold Way'. This is illustrated in this and Box 5.11.

(a) Make two axes: strangeness on the vertical and electrical charge on the horizontal slanted as shown.
(b) Particles with charge $+1$ will lie on the charge $+1$ line. The K^+ has strangeness $+1$ and so occurs at the point common to strangeness $+1$ and charge $+1$. The π^+ has no strangeness.
(c) The K and π particles' positions on the figure yield a hexagonal pattern with a π^0 at the centre.

This similarity in the patterns is very striking. To make them identical would require an eighth meson without strangeness or charge so that it could accompany the π^0 in the centre spot. The discovery of the eta meson (η) in 1961, mass 550 MeV, with no charge or strangeness, completed the pattern perfectly.

Box 5.11   Baryons and the Eightfold Way

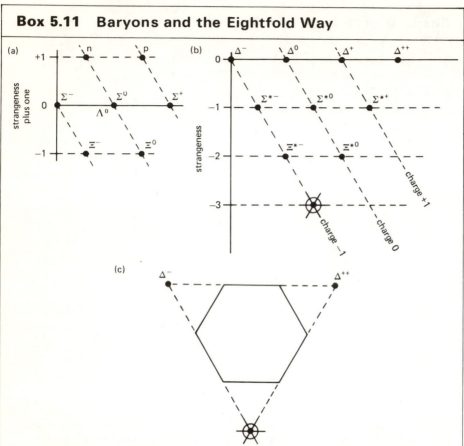

(a) The pattern for the baryons is the same as that for the mesons but for the presence of a second particle at the centre (Λ^0 and Σ^0). The discovery of η^0 at the centre of the meson pattern completed the correspondence.

(b) Heavier baryons were found which were related to the proton, Σ and Ξ baryons. The pattern is the familiar hexagon with extra particles at the top corners. The mathematical theory that was developed to describe these patterns required an inverted triangle to exist containing ten particles.

(c) The $\star$ denotes the position of the particle (Ω^-) required to complete the pattern.

The common pattern for the baryons and for the mesons suggests some important underlying relation between them. These patterns of eights were named the 'Eightfold Way' by Gell-Mann.

We can play the same game with the Δ, $\Sigma^\star$ and $\Xi^\star$ particles because these particles seem to form a family (each decays into analogous members of the octet that contained the proton (Box 5.11 part (a)). Let us put these particles onto a similar figure. We do not find a simple hexagon this time but instead have a hexagon with extra particles at the top corners (the Δ^- and Δ^{++} respectively).

The theory that Gell-Mann and Ne'eman had developed of the Eightfold Way led them to expect that a group of ten should exist.† The pattern in Box 5.11(b) should be completed by extending the hexagon at a third corner, namely the bottom, thereby forming an inverted triangle. The position of the particle that would complete the pattern (which has ten members and is called a decuplet or decimet) would occupy the position indicated by ★ in the figure. It would have strangeness minus three and have negative charge. Gell-Mann named it the omega minus (Ω^-), minus referring to its negative electrical charge and omega (the final letter of the Greek alphabet), in honour of it being the last particle in the pattern, and thus the final step in proving the validity of the scheme.

Furthermore Gell-Mann was able to predict its mass successfully. The Δ particles have zero strangeness and mass 1235 MeV. The $\Sigma^\star$ have strangeness minus one and mass about 1385 MeV, and the $\Xi^\star$ with strangeness minus two have masses of 1530 MeV. Each time you go down the pattern from strangeness zero to minus one and then to minus two the mass increases by about 150 MeV. For this reason Gell-Mann supposed that the particle with strangeness minus three would be another 150 MeV heavier than the strangeness minus two $\Xi^\star$: thus he predicted 1680 MeV for it.

In 1963 at Brookhaven Laboratory in New York, and independently at CERN, Geneva, the predicted particle was found. Its strangeness was minus three, its charge was negative and its mass was 1679 MeV!

Thus the final link in the pattern was found and the relevance of the patterns was established. This pattern is as significant as Mendeleev's table of the atomic elements and in his successful prediction of new elementary particles (such as the Ω^-, Box 5.12) Gell-Mann had paralleled Mendeleev's prediction of the atomic elements gallium, germanium, and scandium.

With the validity of the Eightfold Way established, the crucial question that it posed was—*why*? What is the cause of this pattern underlying the abundance of supposedly elementary particles? The importance of resolving this grew as more and more particles were found (in fact over a hundred species of particle were discovered in the quarter-century following the discovery of the pion back in 1947). It was with the discovery of this multitude that the riddle was solved.

† In Chapter 6 we shall see how these patterns emerge due to an underlying 'quark' structure in nuclear matter.

Box 5.12   Prediction and discovery of the Ω^-

At an international conference on high energy physics held at CERN, Geneva in 1962, the discovery of the $\Sigma^\star$ and $\Xi^\star$ were announced. Yuval Ne'eman, who had discovered the idea of the Eightfold Way independently of Gell-Mann, realised that these particles would fit into a family of ten, and that the tenth member had not yet been seen. Moreover he was able to predict all of its properties. He wrote this all down and gave the note to Gerson Goldhaber, one of the experimentalists involved.

The next day, at the end of a talk about the Eightfold Way the conference chairman called for comments from the floor. Ne'eman raised his hand to make public his discovery and prediction, but the chairman said: 'Professor Gell-Mann'—Murray Gell-Mann had also raised his hand. Gell-Mann went to the blackboard and explained it all, predicting the existence of the Ω^-. Goldhaber inquired afterwards of Ne'eman: 'Did you tell him?'. This was not the first nor the last example of simultaneous and independent discoveries in high-energy physics. (For a more recent example see Chapter 9.)

There is also an amusing postscript. When the Ω^- was finally discovered the experimentalists forgot to tell Ne'eman. He was later sent a nice set of bubble chamber photographs of the discovery with the note 'Please excuse the oversight, but you knew it existed before we did!'

(Source: Y. Ne'eman in *Patterns, Structure and Dynamics* (Univ. of Texas Press, 1981). I am indebted to Prof. Ne'eman for permission to include this anecdote.)

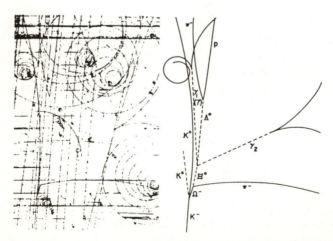

Photo 5F   *The discovery of Ω (Courtesy American Physical Society and Dr Radojicic).*

Box 5.12—continued*

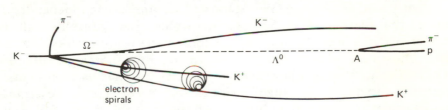

Photo 5G *Bubble chamber photo of Ω⁻ and friends* K⁻ particles produced at CERN were fired into a bubble chamber. On hitting a proton in the chamber an Ω⁻ is produced. This decays into K⁻ and Λ⁰. The Λ⁰ has no charge and so leaves no track, but reveals itself by its decay at A into π⁻ and proton. This characteristic V is illustrated in photo 5C(b). The spirals are electrons that have been knocked out of atoms in the liquid of the bubble chamber. Notice how negative and positive charges bend in opposite directions. (*Courtesy of M. MacDermott et al. of the Universities of Birmingham, Glasgow, Michigan, Paris, and the CERN collaboration*)

6 *Quarks*

Quarks and the Eightfold Way patterns

With hindsight, it is possible to pinpoint Mendeleev's periodic table of the atomic elements formulated in 1869, as the first clue to the presence of a more fundamental layer of matter common to all atoms and responsible for giving them their properties. Half a century later, this was confirmed by the discovery that atoms consisted of electrons encircling a nucleus. By the middle of the present century the structure of the nucleus was in its turn being revealed. The observation of a recurring pattern among the thirty or so nuclear particles known in the early 1960s was an analogous pointer to the possibility of a more fundamental variety of matter—**quarks**—out of which these nuclear particles and ultimately the nucleus are formed.

As early as 1964 Murray Gell-Mann and George Zweig independently noticed that the Eightfold Way patterns would arise naturally if all the nuclear particles were built from just three varieties of quarks. Two of these, known as 'up' and 'down' quarks (u and d for short), are sufficient to build the hadrons that have zero strangeness. Strange hadrons contain the third variety, the 'strange' quark (s for short). The more strange quarks present in a cluster so the more strangeness the cluster has. Refer to Boxes 6.1 and 6.2.

Quarks are very unusual in that they have electrical charges that are fractions of a proton's charge: the up (u) has charge $+\frac{2}{3}$ and the down (d) $-\frac{1}{3}$.

In 1964 no one had seen direct evidence for an isolated body with fractional charge, so Gell-Mann and Zweig's idea received a mixed reception. However there was no denying that the idea worked and although no one has yet prized a quark free from a proton, quarks with these charges have recently been detected *inside* the proton, confirming the hypothesis. If you form a group of three quarks, each one being any of the up, down or strange varieties, then the Eightfold Way pattern of baryons emerges.

Box 6.1 A pair of quarks

If quarks occur in two varieties: up or down, then all of the material in the world about us can be built.

The proton
(Charge +1)

and neutron
(charge zero)

build the nuclei of all atoms. The pions that attract them to one another are formed from up and down quarks and antiquarks.

Box 6.1—*continued*

Up and down quarks and
antiquarks build the pions.

up + antidown: π^+

up + antiup

π^0
contains
both of
these

down + antidown

down + antiup: π^-

Box 6.2   A triangle of quarks

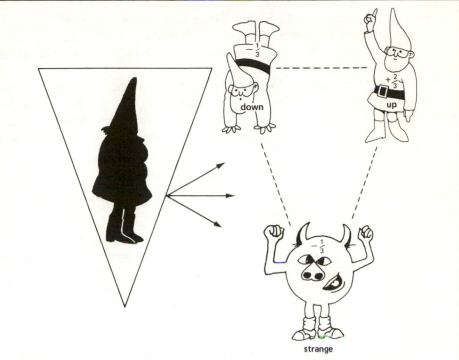

down

up

strange

The existence of strange particles is due to the fact that quarks can occur in three varieties: up, down, or strange.
Quark plus antiquark now yield nine possibilities, and there are strange partners of proton and neutron (e.g. Σ sigma and Λ lambda particles).

Before illustrating how this happens let me answer a question that may have entered your mind: 'Why clusters of *three*—why not two, or five; why not individual quarks?'

These very questions were asked insistently by many physicists in the latter half of the 1960s. All that one could then reply was: 'Because it seems to work that way', and hope that a more complete answer would eventually be forthcoming. Such an answer did subsequently emerge, rationalising all of the assumptions, and will appear later on in the book. I could have chosen to jump forward here to the discoveries of the 1970s which confirmed the reality of quarks, and then backtracked to 1964— presenting Gell-Mann and Zweig's ideas as if they had the benefit of foresight instead of the present approach. If you wish to follow that route, proceed to Chapter 7 first, but that is not how things developed historically. I *am* using the benefit of hindsight to edit out the red herrings and false trails that always plague research at the frontiers, and so I might inadvertently give the impression that progress is inexorably forwards and free from uncertainty. In practice it is not like that. For proponents of the quark theory the latter half of the 1960s was an eerie interregnum when they were able to reach correct conclusions through reasoning that had little or no good foundation, based on weird particles that no one had even seen.

Putting aside this and other justified questions temporarily, suppose that I cluster three quarks in any combination of up, down, or strange. Adding together the electric charges of the quarks gives the total charge of the cluster. Thus two ups and one down will have the same charge as a proton

$$u^{+2/3} + u^{+2/3} + d^{-1/3} = p^{+1}$$

while two down and one up have net zero charge like the neutron

$$d^{-1/3} + d^{-1/3} + u^{+2/3} = n^0$$

Strangeness is a property possessed by strange quarks: the more strange quarks that are present in a cluster so the more strangeness the cluster will have. (The neutron and proton have zero strangeness because they contain no strange quarks.) Furthermore, if the down and up quarks have identical masses and the strange quark is 150 MeV heavier, then one can understand why clusters with a lot of strangeness are heavier than their siblings with less strangeness. The Ω^- with strangeness -3 consists of three strange quarks and is 150 MeV heavier than the $\Xi\star$ (strangeness -2) which is in turn 150 MeV heavier than $\Sigma\star$ (strangeness -1) and this is yet another 150 MeV heavier than the zero strangeness Δ particles.

To see how this all works let us form all possible clusters of three quarks and tabulate the sum of their electrical charges and strangeness using the individual quark properties listed in Box 6.3. We find the following:

Clusters with baryon number = +1	Strangeness = − number of strange quarks	Charge = sum of quark charges	Examples		
uuu		2	Δ^{++}		
uud	0	1	Δ^+	p	
udd		0	Δ^0	n	
ddd		−1	Δ^-		
uus		1	$\Sigma^+\star$	Σ^+	
uds	−1	0	$\Sigma^0\star$	Σ^0	Λ^0
dds		−1	$\Sigma^-\star$	Σ^-	
uss	−2	0	$\Xi^0\star$	Ξ^0	
dss		−1	$\Xi^-\star$	Ξ^-	
sss	−3	−1	Ω^-		

Box 6.3 Quarks

Flavour	Electrical charge	Strangeness
u	$\frac{2}{3}$	0
d	$-\frac{1}{3}$	0
s	$-\frac{1}{3}$	-1
$\bar{u}$	$-\frac{2}{3}$	0
$\bar{d}$	$+\frac{1}{3}$	0
$\bar{s}$	$+\frac{1}{3}$	$+1$

Charge and strangeness of the quarks. The $\bar{u}\bar{d}\bar{s}$ antiquarks have opposite values for charge and strangeness compared to the uds quarks.

The column of ten corresponds exactly with the decuplet of particles that contains the Ω^- (Box 6.4(b)). If we take clusters where at least one quark differs from the other pair then we find the eight members of the octet that contains the proton (that it is eight and not seven is a subtlety arising from the fact that in the (uds) cluster all three are distinct).

The up, down, and strange properties are collectively referred to as the 'flavours' of the quarks. With these three flavours we have readily constructed the Eightfold Way patterns for the baryons. For each and every flavour of quark there is a corresponding antiquark having the same mass and spin as its quark counterpart but possessing opposite sign of strangeness and charge. Thus the strange antiquark, denoted $\bar{s}$, has charge $+\frac{1}{3}$ and strangeness $+1$. By clustering together three antiquarks we obtain the antibaryon counterparts to the octet and decuplet baryons.

With both quarks and antiquarks available, there is another way that we can build clusters with integer charges: form a system containing a single quark and an antiquark. From the three quark flavours and three antiquark flavours in Box 6.3 we can form nine possible combinations $u\bar{d}$, $u\bar{u}$, $d\bar{d}$, $s\bar{s}$, $d\bar{u}$, $u\bar{s}$, $d\bar{s}$, $s\bar{u}$, $s\bar{d}$. The electrical charge and strangeness of each of these is obtained by adding together the quark and antiquark contribution as in the table.

Box 6.4   Baryons

(a) The families of eight and ten baryons, and
(b) the quark systems that generate them.

(a)

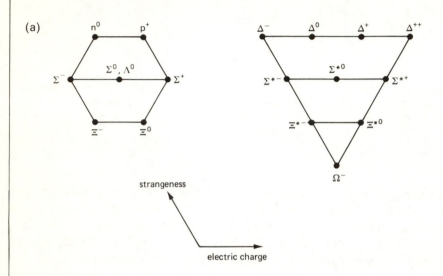

(b)

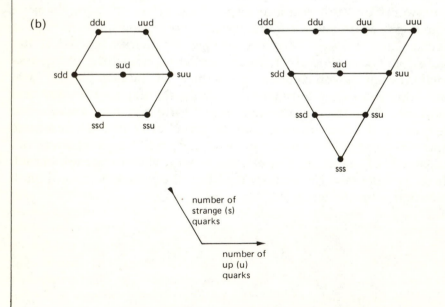

Clusters with zero baryon number	Strangeness = number of strange antiquarks − number of strange quarks	Charge = sum of quark charges	Examples
u$\bar{\text{s}}$	+1	+1	K$^+$
d$\bar{\text{s}}$		0	K^0
u$\bar{\text{d}}$		+1	π^+
u$\bar{\text{u}}$ d$\bar{\text{d}}$ s$\bar{\text{s}}$	0	0	π^0, η^0, η'^0
d$\bar{\text{u}}$		−1	π^-
s$\bar{\text{u}}$	−1	−1	K$^-$
s$\bar{\text{d}}$		0	K^0

Refer to Box 6.5 here.

Box 6.5 Meson nonets

The first hints of the Eightfold Way pattern among the strongly interacting particles (hadrons) came from the family of 8 baryons containing the proton, and a similar hexagonal pattern with 7 mesons (π, K) known in the 1950s. To complete the correspondence between the two, an eighth meson (η) was predicted and its subsequent discovery added much support to the scheme. Later a ninth meson was found (η') which breaks the manifest correspondence. Today we recognise that baryons occur in families of 8 or 10 but mesons occur in nonets (9). This emerges naturally in the quark model where baryons are clusters of three quarks whereas mesons are clusters of quark and antiquark. This is one of the model's many successes.

Astonishingly, these are precisely the combinations of charge and strangeness that mesons are found to have. This is a profound result. For example, there are no mesons with strangeness minus two whereas such baryons do occur, and strangeness minus one states can have charge +1 for baryons but not for mesons. This is precisely what happens in nature and is easily explained by the quark model, as in Box 6.6.

If quarks are real particles then they will have other properties that will be manifested in the hadrons formed from their clusters. For example, it is now known that quarks spin at the same rate as do electrons, namely a magnitude of $\frac{1}{2}$ (in units of Planck's constant $\hbar$). Since an odd number of halves gives a half-integer, then a three quark system has half-integer spin—precisely as observed for baryons. Conversely, an even number of halves gives an integer—quark plus antiquark have integer spins, as do mesons, thus explaining the observation in Box 5.5.

Box 6.6 Quark triangles and the nuclear particle patterns

The triangle occurs if charge is plotted on a skewed axis and strangeness on the vertical axis.

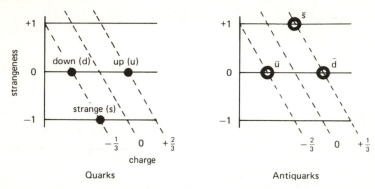

Quarks   Antiquarks

The nine quark plus antiquark possibilities arise if an antiquark triangle is drawn centred on each vertex of a quark triangle.

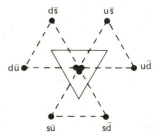

This yields the hexagon pattern for mesons. Three states are at the centre: $u\bar{u}$, $d\bar{d}$, $s\bar{s}$.

The rules for adding up spins (Box 6.7) had been known since the advent of quantum mechanics forty years before these discoveries and had been applied first to electrons in atoms, then later to nucleons in nuclei, and so can confidently be applied to quarks in hadrons, as follows.

Two spin $\frac{1}{2}$ objects, such as quark and antiquark, combine to a total of 0 or 1. Indeed, nine mesons with spin 0 are known—the familiar set containing π and K. Nine spin 1 mesons also exist, consisting of ρ^-, ρ^0, ρ^+ with masses of 770 MeV; and ω^0 of about the same mass; $K\star^-$, $K\star^+$, $K\star^0$, $\overline{K\star^0}$ masses 890 MeV; and the ϕ^0, mass 1020 MeV completes the family nicely. Indeed it was the discovery of these spin 1 mesons, in particular the ϕ in 1963, that played an essential role in the development of the quark hypothesis.

Three spin $\frac{1}{2}$ objects, such as three quarks, combine to a total spin of $\frac{1}{2}$

or $\frac{3}{2}$. The eight members of the family containing the proton each have spin $\frac{1}{2}$; the ten containing the Δ, $\Sigma^\star$, $\Xi^\star$ and Ω^- each have spin $\frac{3}{2}$. The successful explanation of spin = 0 or 1 for mesons built from quark plus antiquark, has been matched by the spin = $\frac{1}{2}$ or $\frac{3}{2}$ for baryons built of three quarks.

An 'atomic' model of hadrons

Atoms consist of electrons in motion about a central nucleus. The nucleus itself has an internal structure consisting of neutrons and protons. The spinning and orbiting motions of the electrons in atoms or the nucleons in nuclei give rise to assorted excited states of atoms and nuclei. Thus if hadrons are clusters of quarks we should expect by analogy that excited hadronic states will occur as a result of the various spinning and orbiting motions of their constituent quarks.

When the quarks have no orbital motion about one another then the total spin of the cluster comes entirely from the spins of the individual quarks within. We have already seen how this gives spin 0 or 1 mesons and spin $\frac{1}{2}$ or $\frac{3}{2}$ baryons. What happens if we now admit orbital motion for those quarks in addition to their intrinsic spins?

The total spin of the hadronic cluster will result from the quarks' spins and also their mutual orbital angular momenta. The more orbital motion the quarks have, so the larger the total spin of the hadron will be. Quarks in a state of rapid orbital motion carry more energy than when orbiting slowly, thus the energy or mass of a hadron with large spin will tend to be higher than that of hadrons with small spins.

This is indeed the case in nature. Hundreds of hadrons have been discovered during the last 20 years, and the higher their spin so the larger their masses are seen to be. There are so many that no one can memorise all of their properties—these are listed and revised biennially in a publication of increasing bulk. Fortunately we do not have to know them all. Fermi is reputed to have said that if it were necessary to know the names of all the hadrons then he might as well have been a botanist. We can easily summarise them though. These hundreds of particles form the ubiquitous hexagonal families of the Eightfold Way, the spin 0 and 1 mesons being accompanied by spin 2, 3, and 4 mesons (Box 6.8). Baryon patterns have been observed with spin $\frac{1}{2}$, $\frac{3}{2}$, $\frac{5}{2}$ and so on up to $\frac{15}{2}$ (so far!). The Eightfold Way patterns bear testimony to the quark constituents, the increasing spins and masses exhibiting the dynamic motion of those quarks within the hadrons.

The picture we have today is that hadrons are clusters of quarks much as atoms are clusters of electrons and nuclei. There are tantalising similarities here, but also some profound differences.

Box 6.7   Angular momentum

When adding together two or more angular momenta we must take account of their vector character: the direction of spin is important in addition to its magnitude.

In subatomic systems, the angular momentum is constrained to be an integer multiple of Planck's quantum $\hbar$:

$$L = n\hbar \qquad (n = 0, 1, 2 \ldots; \text{ known as S, P, D} \ldots \text{states})$$

The sum of two angular momenta must itself be an integer multiple of $\hbar$. Thus, depending upon the relative orientations of L_1 and L_2 the sum can have any value from $(n_2 + n_1); (n_2 + n_1 - 1) \ldots$ to $|n_2 - n_1|$. (Each of these is understood to be multiplied by $\hbar$ but conventionally the $\hbar$ is often omitted and so we write $L = 3$ not $3\hbar$.)

Electrons have an intrinsic angular momentum or 'spin' of magnitude $\tfrac{1}{2}\hbar$. Adding to $L = n_1\hbar$ gives $(n_1 + \tfrac{1}{2})\hbar$ or $(n_1 - \tfrac{1}{2})\hbar$. The difference of these is an integer multiple of $\hbar$. In general adding $n_1\hbar$ to $\dfrac{m}{2}\hbar$ (m odd or even) gives:

$$\left(n_1 + \frac{m}{2}\right); \left(n_1 + \frac{m}{2} - 1\right) \ldots \left|\left(n_1 - \frac{m}{2}\right)\right|$$

as the set of possibilities.

Some specific examples may help make the point. Two spin $\tfrac{1}{2}$ will add to either 1 or 0. Three spin $\tfrac{1}{2}$ will yield $\tfrac{3}{2}$ of $\tfrac{1}{2}$. Two $L = 1$ will add to yield total 2, 1, or 0.

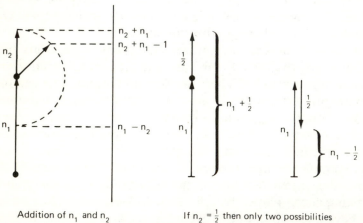

Addition of n_1 and n_2 If $n_2 = \tfrac{1}{2}$ then only two possibilities

When energy is supplied to a hydrogen atom, the electron is raised into states of higher angular momentum and energy. If enough energy is supplied the electron will be ejected from the atom and 'ionisation' occurs. The energy to excite electrons from the ground state to higher energy states is of the order of a few electronvolts.

Box 6.8   More Eightfold Way hexagons

When particles with the same spin and other properties were collected together, the ubiquitous hexagonal patterns were found over and over again. This is illustrated here for some of the mesons. Where there are blank spaces in the patterns no meson has yet been found to fit in the slot. Given the validity of the patterns, one can confidently predict that mesons must exist with the properties required to fit into the blank positions in the figures. The $^3S_1$ etc. denote the spin and orbital states of the quark and antiquark forming these mesons.

Compare this with quark clusters. To excite a quark requires hundreds of MeV (the S-state mesons π and ρ weigh 140–770 MeV, their P-state counterparts weigh some 1000–1300 MeV). This is in part due to the fact that mesons are much smaller than atoms and so typical energies are correspondingly greater (the uncertainty principle underwrites microscopic phenomena—small distances correspond to high momentum or energy and vice versa). It is also due to the nature of the forces amongst quarks which are much stronger than electromagnetic forces and so provide more resistance to excitation.

The other noticeable feature is that although the patterns of increasing energy with increasing spin are essentially the same in quark clusters and atoms, the relative energy gaps between analogous configurations in the two cases are quite different (Box 6.9). In hydrogen, the amount of energy required to excite the electron from the S to P configuration is already nearly enough to eject it right out of the atom. In quark clusters things are not like this. The separation of S to P configurations is roughly the same as P to D and so on. As energy is supplied to a quark cluster the quarks are excited to a higher energy configuration but are not ejected from the cluster—there is no known analogue of ionisation.

Box 6.9 Pattern of energy levels in atoms and quark clusters

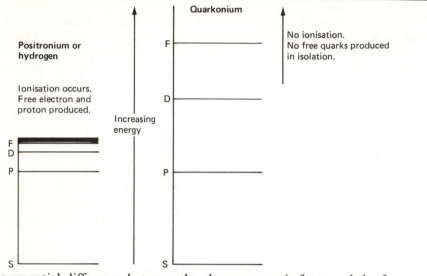

The essential difference between the electromagnetic force and the force that binds quarks is that electric charges can be liberated from their binding in atoms (e.g. by heat) whereas quarks seem to be permanently confined inside their 'atomic clusters' (hadrons). Apart from this, the pattern of levels, correlation of increasing mass and spin, splitting of masses for parallel from antiparallel spin are all very similar.

Quarks in the proton

By the late 1960s more than a hundred species of subnuclear particles had been found. Each and every one of these had properties suggesting that it was built from quarks as in the above scheme. The accumulating evidence convinced most physicists that the quark hypothesis was probably correct even though individual quarks had not been found; quarks had only manifested themselves in their clusters—the subnuclear particles.

Many searches for isolated quarks have been performed during the last 15 years and yet no one has conclusively observed a single quark. If you believe that quarks exist inside subnuclear particles then their refusal to expose themselves to such intense searches is a puzzle. Their most dramatic property is that they are required to have electrical charges of $\frac{2}{3}$ and $-\frac{1}{3}$ (Box 6.3) whereas all particles ever seen and confirmed have integer or zero charges. This possession of a fractional electric charge should make an isolated quark very obvious. It was suggested by some that quarks might be so massive that no accelerator on Earth was powerful enough to produce them. However one would expect them to be produced in collisions between cosmic rays and the atmosphere and even if this happened only rarely, their fractional charges would be so distinctive that it would be hard not to notice them.

With the failure of the early quark searches, suspicion grew that quarks were not really physical objects but were somehow mathematical artefacts which conveniently produced the patterns of the particles while having no *real* significance in themselves.

The turning point came as a result of a series of experiments at SLAC Stanford, California from 1968 and at CERN, Geneva from 1970 which saw direct evidence for quarks physically trapped inside the proton and neutron. The significance of these experiments parallels Rutherford's 1911 discovery of the atomic nucleus and they were in essence simply more powerful versions of his experiment. Indeed there are many similarities in the revelation of atomic substructure at the start of the century and the uncovering of the quark layer of matter during the last 15 years.

Early in the 1950s collisions of protons with other protons had shown that they had a diameter of about 10^{-13} cm, very small compared to the size of a nucleus but more than a hundred times bigger than an electron. With this discovery that the proton was 'large' suspicion arose that it might have an internal structure.

The original guess was that the proton's size was due to a cloud of pions perpetually surrounding it. As a result the proton was pregnant with the carriers of the nuclear force by which neighbouring protons or neutrons could be ensnared and nuclei formed. Although appealing, this failed to give a quantitative description of nucleon properties such as magnetic moments.

After the quark hypothesis first appeared in 1964, the idea gained ground that quarks in motion gave the proton its size, perhaps in analogy to the way that electrons and nuclei gave size to atoms.

Early in the 1960s a two-mile long machine was built at Stanford in California capable of accelerating *electrons* until they had energies in excess of 20 GeV (Photos 6A and 6B). At these high energies electrons can resolve structures less than 1 fermi in size and are therefore a perfect tool for probing inside protons and investigating their structure.

The electron's negative charge causes it to be attracted or repelled respectively by the up and down quarks which have electrical charges $+\frac{2}{3}$ and $-\frac{1}{3}$. The quarks' spinning motion causes them to act as magnets which exert calculable magnetic forces on the passing electrons. Taking all of these things into account it is possible to predict what should happen when an electron beam hits a proton at high energy. You can calculate the chance that it is scattered through some angle, how much energy it loses while the proton recoils, and so on.

By studying the way that the electron is scattered by the proton you can determine where the charge of the proton is concentrated. If it was evenly distributed throughout the whole volume then the proton would be seen as a diffuse cloud of electricity and the electron beam would pass through with little or no deflection. However, if the charge is localised on three quarks then the electron will occasionally pass close to a concentration of charge and be violently deflected from its path, analogous to the old alpha particle experiments that revealed the nuclear atom in 1911.

Violent collisions *were* seen at SLAC, California and the distribution of the scattered electrons showed that the proton was indeed built from entities with spin $\frac{1}{2}$ such as quarks. Comparison of these results with similar experiments at CERN in Geneva (where neutrinos were used as probes in place of electrons) showed that these constituents have electrical charges which are $\frac{2}{3}$ and $-\frac{1}{3}$ of the proton's charge. These are identical to the properties that had been deduced for quarks from the observed Eightfold Way patterns, and confirm the quarks' presence in the proton.

Some additional discoveries about the inside of the proton were made. First of all, the experiments showed that some electrically neutral particles ('gluons') exist there in addition to quarks. Just as photons are the carriers of the electromagnetic force between electrically charged particles, so it was suspected that these gluons might be carriers of the force that attracts quarks together forming the proton. More recent discoveries have confirmed the existence of gluons and have identified a new property, 'colour', carried by quarks (but not by electrons) which is similar to electrical charge (Chapter 7): as photons are to electrical charge so are gluons to colour.

The discovery of gluons is most welcome as it provides a strong hint that we are indeed revealing the inner workings of the proton, not just what it is made from but how it is held together. However there were other discoveries which were paradoxical and ultimately of the greatest significance. First, quarks appear to be very light, less than one-third of a proton mass. Second, they appear to be almost free inside the proton as if they are hardly glued together at all!

If this is really what quarks are like then you would expect them to be easily ejected from the proton. Indeed, as soon as these phenomena were seen in 1968, plans were made to see what happened to the proton after it had been struck so violently. For a year or so there were hopes that individual quarks might emerge, but these hopes were short lived—pions and other familiar particles were produced but no free quarks or gluons appeared.

Photo 6A *SLAC: 2 mile accelerator of electrons* Stanford Linear Accelerator of electrons (SLAC) in California. This is like a gigantic television. The back of the tube is at A. Electrons accelerate along the tube which is two miles long and the 'screen' consists of detectors in the building at B. (*Courtesy Stanford University*)

Photo 6B *Electron detectors at SLAC* (*Hall B in photo 6A*) These are the detectors referred to in photo 6A. A target of protons sits at the extreme left of the picture and electrons enter from the left, are scattered, and then detected. (*Courtesy Stanford University*)

Box 6.10 The quark force paradox (late 1960s)

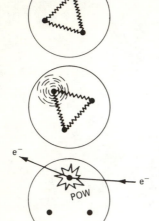

The proton can be viewed as three quarks very tightly bound by super-strong force.

Exciting one or more quarks yields baryon resonance states like Δ.

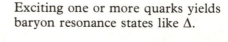

SLAC, California (Photo 6A). High energy electrons scatter from quarks in the proton, giving the first direct evidence for quarks. Paradox: quarks appear to be free!

Neutrinos scatter from protons at CERN. Comparison with electron scattering reveals that in addition to the quarks there is also electrically neutral material inside the proton, dubbed 'gluons'. Suspicion arises that this may be the glue that holds the quarks to one another in the proton.

1970s Proton: quarks held by gluons

Paradox: Gluons bind quarks very weakly; but try to get a quark out and the glue comes on strong.

Theorists search for a theory of quark force that will have these properties (see Chapter 7).

Although this was a disappointment for the experimentalists it created an exciting paradox for the theorists (Box 6.10). As one physicist succinctly put it, 'The proton is like an ideal prison: the quarks have perfect freedom inside but are unable to escape'. It is analogous to quarks being held to one another by an elastic band that is slack. The quarks are apparently free but after being struck they recoil and the elastic becomes tighter, preventing them from escaping. The elastic may become so stretched that it snaps. The two new ends created have quarks on them, and so mesons are formed but not free quarks.

This is the picture we now have of the quarks in the proton but no one has yet managed to explain fully how gluons cause the quarks to cluster in this way. This paradox was seminal in the subsequent development of theoretical physics and a theory of the quark forces has been constructed in recent years that does appear to have all of the desirable properties of free, permanently bound quarks. This theory is known as 'Quantum Chromodynamics' or QCD and its development and present status are the next topics in this story.

Before proceeding let me answer a question that might have occurred to you. Does the notion of colour and its role in generating forces among the quarks mean that we now have *five* fundamental forces—gravity, weak, electromagnetic, strong nuclear force, and this new quark (colour) force?

In fact we still have only four: the strong nuclear force between neutrons and protons is now believed to be a complicated manifestation of the more fundamental colour force acting between their constituents—quarks. It may be helpful to draw a historical analogy. In the early 19th century intramolecular forces were thought to be fundamental. We now realise that they are but complicated manifestations of the more fundamental electromagnetic force acting on the atomic constituents—the electrons. As we shall see, there exists a profound parallel between

electric and **colour** forces

acting on

electrons and **quarks**

which are the constituents of

atoms and **protons/neutrons.**

These in turn form

molecules and **nuclei.**

The historically identified

molecular and **strong nuclear** forces

are manifestations of the fundamental

Nature does indeed appear to be efficient, not just in the fundamental particles (leptons and quarks) but in the forces that bind them to form bulk matter.

7 Quantum Chromodynamics: A theory for quarks

Colour

There are some interesting parallels between the discovery that nucleons are built from quarks and the earlier discoveries of atomic and nuclear substructures. There are four common ingredients whose psychological impacts have been rather different in each case, primarily because the sequence in which they came to the fore differed.

1 Objects that are supposedly independent of one another and are structureless, nevertheless exhibit common properties. These features are distributed or repeated among the objects in a regular fashion such that a definite pattern is seen (e.g. the periodic table of atomic elements or Eightfold Way of elementary hadrons). This gives a hint that supposedly structureless objects might in fact have common constituents.

2 When particles beams (such as electrons, neutrinos, or alpha particles) are fired at the objects of interest, the beams are scattered. The violence of this scattering shows that the target is not diffuse but contains a complex inner structure. This proves that the 'structureless objects' are built from more fundamental constituents.

3 Attractive forces are identified which bind the constituents to one another thereby forming the more complex structures.

4 Pauli's exclusion principle plays a crucial role, governing the behaviour of the constituents. This in turn limits the number and form of the complex structures such that the allowed ones exhibit common properties and patterns of regularity. Pauli's exclusion principle is most familiar in atoms, where it forbids more than one electron occupying the same energy state and is thereby responsible for generating the regular pattern of atom varieties discerned by Mendeleev. Pauli's principle applies also to quarks and should control the ways that they can combine to form hadrons and generate the Eightfold Way patterns.

The electron and nuclear structures of the elements were identified long after the electromagnetic force had been understood. Thus the forces at work were already known even though their role in generating

Box 7.1 **A summary of our knowledge of atoms, nuclei and hadrons as it might have appeared in the late 1960s**

	Atom	*Nucleus*	*Hadron*
Pattern	Mendeleev table 1869	Isotopes and magic numbers known early 20th century	Eightfold Way 1962
Constituents identified	α-particle scattering	α-particle scattering	electron and neutrino scattering
	⇒ nucleus 1911 ionisation ⇒ electrons	⇒ proton 1919 . . . neutron 1932	⇒ quarks 1968–70
Clustering force	Electromagnetic force (already known)	Strong force (inferred 1935)	? [Unknown]
Force carrier and theory	Photon	Pion	
	Quantum electro- dynamics 1948	Yukawa model	? [Unknown]
Pauli principle examples	Electrons occupy energy 'shells'	At most two protons and two neutrons in lowest energy state	Forbids three identical strange quarks to simultaneously occupy lowest energy state
	⇒ chemical regularity	⇒ α-particle stable	
	⇒ Mendeleev table	⇒ isotopes	⇒ Ω⁻ can not exist ? [Paradox]

atomic structure was not fully understood until quantum mechanics was formulated around 1928.

For quarks and elementary particles the historical order was reversed from the above case.

Ingredients 1 and 2 had been recognised in the 1960s with the Eightfold Way patterns of hadrons, and the electron and neutrino scatterings from the quarks within those hadrons. Very little was known about the forces that act on quarks beyond the fact that quarks cluster in threes (baryons) or quark with antiquark (mesons) and in no other manner. This is an empirical fact which must be accounted for by any theory of quark forces but in the absence of other clues is not sufficient to lead us to that theory. The first real clue, though it was not recognised as such at the time, was a paradox concerning the Pauli exclusion principle.

Spin $\frac{1}{2}$ particles (such as electrons, protons, neutrons, and quarks) obey this principle which forbids more than one of a given kind being in the same state of energy and spin. A familiar example is in the formation of atoms, where Pauli's principle forces electrons into particular orbital configurations and as a result a periodically repeating pattern of chemical properties occurs, as originally noted by Mendeleev. In nuclear physics the principle allows at most two protons and two neutrons in the lowest energy state—this is the source of the stability of the α-particle, helium–4.

Quarks have spin $\frac{1}{2}$ and so the principle should apply to them too: two or more quarks cannot occupy the same state if they have identical flavours. It is natural to expect that the lightest clusters are formed when each quark is in its lowest energy state, thus the Ω^- which consists of *three* identical strange quarks is seemingly forbidden to exist, contrary to clear evidence that it does!

Box 7.2   The Ω^- problem and the need for coloured quarks

One strange quark spinning clockwise.

Second strange quark. It must spin the opposite way so that it is distinguishable from the first.

Third quark has only two possible ways to spin—clockwise or anticlockwise. But both are forbidden as there are already quarks present in these states.

Box 7.2—*continued*

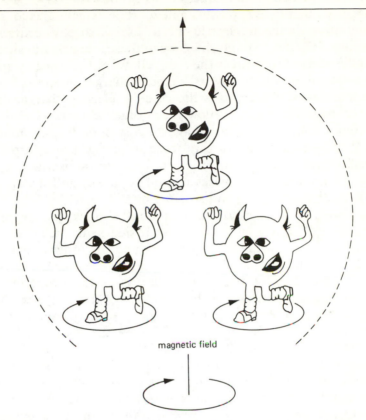

magnetic field

Ω^- is seen to exist and spin in a magnetic field with each quark spinning clockwise.

How can this be?

Quarks must possess some further property that enables them to be distinguishable.

If strange quark can exist in any of three colours we can distinguish each one in the Ω^-.

Oscar (Wally) Greenberg recognised this problem with the Pauli principle soon after the idea of quarks was proposed in 1964. To resolve it, he suggested that quarks possess a new property called 'colour' which is in many ways similar to electric charge except that it occurs in three varieties (Box 7.2). To distinguish between these, particle physicists have whimsically referred to them as the red, yellow, or blue variety and these charges are known collectively as 'colour' charges. Instead of simply positive or negative charge, as is the case for electric charge, there are positive or negative 'red', 'yellow', or 'blue' colours. Quarks carry positive colour charges, antiquarks have the corresponding negative colour charges. Thus a strange quark can occur in any of these forms and to distinguish them we append the appropriate subscripts s_R, s_Y, s_B. Similarly the up and down quarks can be u_R, u_Y, u_B and d_R, d_Y, d_B. For antiquarks we have $\bar{s}_R, \bar{s}_Y, \bar{s}_B$ and so on (Box 7.3).

Pauli's principle only forbids *identical* quarks to occupy the same spin and energy state. Thus if one of the strange quarks in the Ω^- carries the red variety of charge, while one has yellow and the other one blue, then they are not identical and so the Ω^- can exist—as empirically observed.

Given that before 1968 the quark idea was by no means universally accepted, it is quite remarkable that Greenberg did not take the Pauli paradox as evidence that the quark model was wrong, but instead proposed the property of colour to overcome the problem. The chief merit of the quark model in 1964 was that it provided a *simplification*; his idea of introducing a new property—which in effect multiplied the number of quarks by three—was not pursued with much enthusiasm at first. The quark model began to be taken seriously once the quarks were seen inside the proton in 1968 (p. 83). Although these experiments showed that quarks were present, they were incapable of showing whether or not quarks had colour. The crucial evidence for coloured quarks has centred on experiments where electrons collide with their antimatter, positrons, and mutually annihilate.

When an electron and positron annihilate, the energy of their motion is converted into new varieties of matter such as a muon and antimuon or quarks and antiquarks. The quarks and antiquarks cluster together forming the familiar nuclear particles such as protons and pions and it is these that are detected. The probability that nuclear particles emerge relative to the probability that muons and antimuons emerge, is given by the sum of the charges squared of all varieties of quarks that are confined inside those nuclear particles. The relative abundances are predicted to be

$$\frac{\text{Production rate of nuclear particles}}{\text{Production rate of muon and antimuon}} = [\underbrace{(\tfrac{2}{3})^2}_{\text{up}} + \underbrace{(-\tfrac{1}{3})^2}_{\text{down}} + \underbrace{(-\tfrac{1}{3})^2}_{\text{strange}}] \times \underbrace{3}_{\substack{\text{if three} \\ \text{colours}}}$$

Box 7.3 Colour, the Pauli principle, and baryon spins

The Pauli principle forbids two identical spin $\frac{1}{2}$ particles to occupy the same state of energy and spin. Thus the two electrons in the ground state of the helium atom must spin in opposite directions—antiparallel. Similarly the two protons in an α-particle must spin antiparallel as must the two neutrons.

The same would be true for quarks if they did not possess colour which distinguishes the otherwise identical strange quarks in the Ω^- or the two up quarks in the proton for example. The effect of colour combined with the Pauli exclusion principle is that any two quarks having the same flavour (two up, two down, two strange) in the lowest energy state must spin *parallel*—precisely the opposite of what happens in atoms and nuclei.

An extreme example is when all three quarks have the same flavour, as in Δ^{++}(uuu), Δ^-(ddd) or Ω^-(sss). Here all three quarks must spin parallel, hence the total spin is $\frac{3}{2}$ (Box 6.7, part (b)).

If two flavours are identical and the third differs (e.g. Δ^+(uud) or p (uud)) then the identical pair must spin parallel but the third is not constrained, it can spin parallel (hence total spin $\frac{3}{2}$ as in the Δ) or antiparallel (hence total spin $\frac{1}{2}$ as in the proton).

Thus we see that the decuplet containing Δ^{++}, Δ^-, Ω^- naturally has spin $\frac{3}{2}$ as observed. Removing the cases where all three quarks are identical leads to the octet where the total spin is $\frac{1}{2}$ precisely as in nature. When all three quarks have different flavours (uds) then any pair can be spinning parallel or antiparallel, hence the extra state at the uds site in Box 6.4. The Σ^{*0} has the ud and s all parallel; the Σ^0 has ud parallel and opposite to s while the Λ^0 has ud antiparallel.

Identical quarks spin parallel

$$\Delta^{++} \quad S = \tfrac{3}{2}$$

Box 7.3—*continued*

Two identical
spin parallel   and   One unique can
spin eitherway

+

Δ^+ $S = \frac{3}{2}$

or

+

p $S = \frac{1}{2}$

Three different quarks
Up–down parallel make Σ type

$\Sigma^{\star 0}$ $S = \frac{3}{2}$

Box 7.3—*continued*

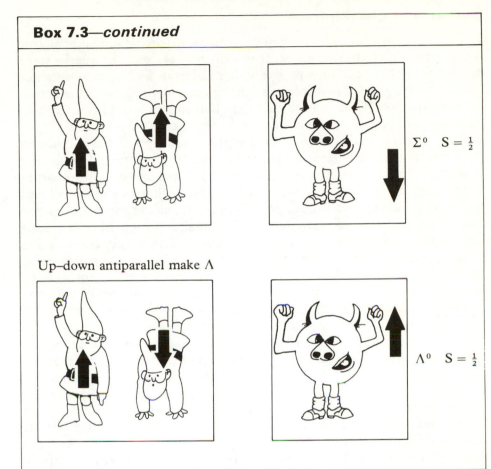

Up–down antiparallel make Λ

Σ^0 $S = \frac{1}{2}$

Λ^0 $S = \frac{1}{2}$

hence the result is $\frac{2}{3}$ if quarks do not have colour (like leptons) but 2 if they occur in three colours. This experiment was performed in 1970 at Frascati near Rome. The ratio was seen to be much larger than $\frac{2}{3}$ and consistent with 2 within experimental uncertainties (though to be historically accurate I should stress that the experimental uncertainties were rather large. It was not until these experiments were reproduced with greater precision at Stanford, California after 1972 that evidence for colour started to become convincing). At last, here was direct evidence supporting the notion that there are three distinct colours of up quark, three down, and three strange.

Colour and quark forces

Quarks, and hadrons containing quarks, all experience the strong nuclear force, whereas electrons and neutrinos do not. As soon as quarks were discovered to have colour, a property that electrons and neutrinos do not have, it was suggested that colour might be the source of the forces acting between quarks. If this was correct then naturally it would explain why electrons and neutrinos are blind to the strong nuclear forces.

How shall we build a theory of colour? The inspired guess made in 1972 was that colour interactions are analogous to the interactions among electric charges. In electrostatics, like charges repel and opposite charges attract and the analogy for colour is immediate—like colours repel and opposite colours attract. Thus two red quarks repel one another but a red quark and an 'anti-red' antiquark will mutually attract. Similarly blue attracts anti-blue or yellow attracts anti-yellow. This is very encouraging because it explains the existence of mesons naturally—just as positive and negative electrical charges attract to form net uncharged atoms so have positive and negative colours, carried by quark and antiquark, attracted to form net uncoloured hadrons.

In electrostatics, two positive charges are always 'like charges' and repel. For colour, two red quarks are always coloured alike and repel, but what about a red quark and a blue quark? These are alike in that they are both quarks ('positive colours') but unlike in that the colours differ.

It turns out that these different colours can attract one another but less intensely than do the opposite colours of quark and antiquark. Thus a red quark and a blue quark can mutually attract and the attraction is maximised if in turn they cluster with a yellow quark (Box 7.4). Red and yellow, red and blue, blue and yellow all attract one another and thus do the three quark clusters known as baryons occur. Notice that the baryon formed this way necessarily contains three quarks *each one with a different colour*. Thus we have been led naturally to the picture that Greenberg invented *ad hoc* in 1964 as explanation of the Pauli exclusion paradox.

The mathematics that has been developed to describe colour interactions shows that the above clustering—quark and antiquark of opposite colours or three quarks each of different colours—are the only ways that net uncoloured hadrons can be formed. Nature seems only to allow uncoloured systems to exist free of one another; colour is confined in clusters where the net colour cancels out. We do not yet know whether this is absolutely true or whether there is instead a small chance for colour-carrying particles to be produced in the laboratory.

Box 7.4   Colour attractions form mesons and baryons

Colour and opposite colour attract. Thus is formed a **meson** = $q\bar{q}$.

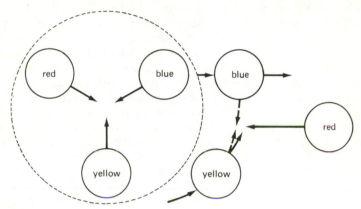

Three different colours attract. Nearby like colours are repelled. Colourless clusters of three different coloured quarks form. Hence **baryons** = $q_R q_B q_Y$.

Gluons

Combining electrostatics with relativity and quantum theory generated the successful theory known as quantum electrodynamics (QED). The idea behind the quantum chromodynamic theory (QCD) of quark forces is that colour generates them in the same way that electric charge generates electromagnetic forces. Mathematically, as quantum electrodynamics is to electric charge so quantum chromodynamics will be for colour.

In quantum electrodynamics, photons necessarily exist. They are the massless, spin 1 carriers of the electromagnetic force between electrically charged objects. In quantum chromodynamics the analogous requirement is that 'gluons' exist—these are massless spin 1 carriers of the force between coloured objects. See Box 7.5.

In quantum *electro*dynamics acceleration of an electric charge leads to radiation of photons. Analogously in quantum *chromo*dynamics acceleration of colour radiates gluons (Box 7.6). Both of these phenomena occur in the electron–positron annihilation process if quarks are produced. The electron (e^-) and the positron (e^+) annihilate and produce a photon (γ)

Box 7.5   Gluons

Gluons carry colour and can interact with one another by exchanging further gluons. This is in marked contrast with QED where the photon is electrically neutral and so does not self-interact. This difference can be traced back to the fact that QCD is a theory of more than one colour and so a particle might change its colour from one to another type (say red to blue) while emitting a gluon—which will possess the 'colour mismatch'—(say purple). In QED by contrast there is only one charge and this is retained by the particle emitting an electrically neutral photon.

The only essential difference between QCD and QED is the threefold colour as against unique electric charge. This small difference causes the force carriers to carry colour and so to interact with one another even while transmitting the force from one point in space to another. Photons, in contrast, do not interfere with each other during the journey. As a result the way that the force acts across space differs from electromagnetism. And so from such a trivial modification far reaching consequences emerge—the force which is quite weak transmitted over a short distance grows dramatically at large distances. The consequence that most directly affects us is that after a distance of 10^{-13} cm the force becomes very strong—so strong that it will hold two protons to one another overcoming their electromagnetic repulsion. So nuclei are formed at the heart of the atoms.

Gluons are confined in clusters for the same reasons that quarks are—they carry colour and so are constrained by the same force as the quarks. As a result we can no more study isolated gluons than isolated quarks and have to seek indirect manifestations of their existence.

If quarks cluster to annul their net colour, forming white systems (protons, pions . . .) so might gluons cluster forming white 'glueballs'. These are currently being looked for.

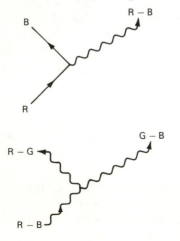

A red quark turns into blue by emitting a gluon whose colour is 'red minus blue'. Thus the gluon carries colour itself.

A 'red minus blue' gluon converts into a 'red minus green' by emitting a 'green minus blue' gluon. As we proceed up the diagram the net 'red minus blue' is always preserved. In this way a single gluon can fragment into two—something that has no analogue for photons. This causes colour and electric fields to behave differently at large distances (see Box 7.7).

Box 7.6 Quarks and gluons in electron–positron annihilation

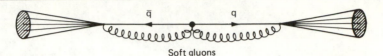

Soft gluons

Two jets of hadrons produced, one on each side. (1975 on)

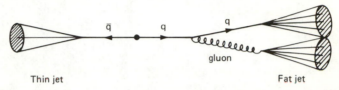

Thin jet   Fat jet

QCD: quark emits gluons. If there is a small angle between quark and gluon then their separate jets are not resolved. Thin jet on one side, fat jet on other side. (TASSO group, 1979)

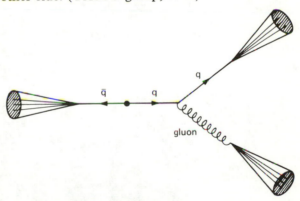

QCD: If there is a larger angle between quark and gluon there are three distinct jets, one each from quark, antiquark, and gluon. (1979)

which then converts into a quark (q) and antiquark (q̄). This sequence of events is conventionally written:

$$e^+ + e^- \rightarrow \gamma \rightarrow q + \bar{q}$$

But the quark and antiquark carry both colour and electrical charge and in the act of being produced the quarks radiate gluons and photons. So the real process is

$$e^- + e^+ \rightarrow \gamma \rightarrow (q + \text{photons} + \text{gluons}) + (\bar{q} + \text{photons} + \text{gluons})$$

(a)

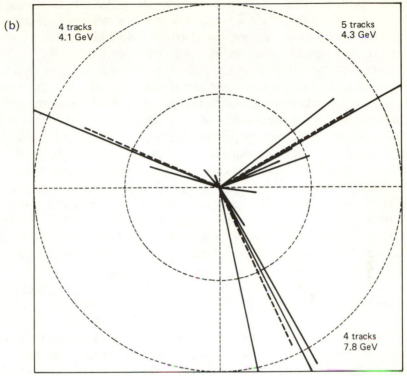

(b)

4 tracks
4.1 GeV

5 tracks
4.3 GeV

4 tracks
7.8 GeV

Photo 7A (*a*) *The TASSO detector at Hamburg's electron–positron storage ring* Electrons and positrons can annihilate in Hamburg's 'PETRA' machine with more violence than in Stanford's SPEAR. The TASSO detector is illustrated. Detection of the upsilon particle and possible evidence for gluons have recently been found at PETRA by this and similar detectors. (*Courtesy TASSO group*)

(*b*) *Evidence for gluons?* Three jets of particle tracks are in this event, seen in the TASSO detector at Hamburg. The dotted lines indicate the direction that the parent quarks and gluon were moving before transmuting into hadrons. It is not yet possible to determine which tracks are from the gluon and which from the quarks.

From established QED one can calculate how much of what is observed is due to the photon radiation. You can then study what is left over and seek characteristics associated with gluon radiation from coloured quarks and compare the resulting phenomena with the QCD predictions.

It would be easy to test the theory if we could directly detect the quarks and gluons created by the electron–positron annihilation. However nature is not so kind, only conventional pions, protons and other clusters of quarks or gluons appear, isolated quarks and gluons do not emerge. The following scenario is predicted by QCD theory and is probably indeed what takes place, but more work is needed before we can be totally sure of its validity. The quark and antiquark move off in opposite directions. Initially they feel no force but as they separate, the force between them grows. The energy in the force field between them is so great that further quarks and antiquarks are spontaneously created. These cluster together so quickly forming mesons and baryons that the original quark and antiquark are not detected, instead two oppositely directed showers of nuclear particles emerge. The quark and antiquark are like abominable snowmen—they are far gone, their footprints—the two jets of nuclear particles—are all that remain. By studying these jets the properties of their parents—quarks or gluons—may be inferred.

QCD predicts that at high energies the quark and antiquark usually carry off most of the energy with the glue collimated along the q$\bar{\text{q}}$ axis, carrying little energy itself. In such circumstances two distinct jets of particles emerge.

Many events with this 'two-jet' characteristic have been seen and studied since 1975. In some experiments the annihilation involved low energy electrons, in others the electrons had very high energies. At low energies the jets are smeared out, distributed about the direction of the parent quarks' motion. QCD predicts that at higher energies the jets should become increasingly collimated and the data do indeed seem to support this. Although these 'two-jet' events dominate the data, there is a chance that the quark (or antiquark) radiates a gluon of large momentum which will deflect the quark (antiquark) from its path (Box 7.6 parts (b), (c)). If the angle between the gluon and quark is small then it will not be possible to distinguish the hadrons coming from each; there will be a thin jet on one side and a fat jet on the other. The first evidence for such events was announced in mid-1979 by the TASSO collaboration (Photo 7A) working at Hamburg.

Sometimes, albeit more rarely, the deflection of the quark when it emits a gluon will be large enough for the individual jets to be resolved. This will yield three jets of particles. Events of this type were discovered during the summer of 1979. Subsequent experimental studies of these three jets have verified that two are emerging from spin $\frac{1}{2}$ quarks while the third originates in a spin 1 object—the gluon? This is as predicted by quantum chromodynamics, though more experimental work remains to be done before we can claim to have 'verified' the theory.

Quantum chromodynamics and electrodynamics

If QCD is mathematically so similar to QED, then we might expect to find similar behaviours for the forces between quarks in clusters (qqq like the proton or q$\bar{\text{q}}$ like the pion) and the electromagnetic forces between the electrons and nuclei of atoms. Such similarities are indeed observed, most noticeably in the hyperfine splittings between certain energy levels of the atom or quark clusters. In hydrogen there is a magnetic interaction between the electron and proton which contributes a positive or negative amount to the total energy depending upon whether the electron and proton are spinning parallel or antiparallel (total spin 1 or 0). In quark clusters there is an analogous 'chromomagnetic' interaction between pairs of quarks which adds to the energy of a spin 1 pair, and depletes that of a spin 0 pair. Such a splitting is indeed seen for quark–antiquark systems (mesons) for example where the $^3S_1$ combinations (ρ, $K^\star$, ϕ) are some hundreds of MeV heavier than their $^1S_0$ counterparts (π, K, η).

Not only does this behaviour illustrate the similarity between QCD for quark clusters and QED for atoms, but it also shows the relative strengths of the two forces. In atoms these hyperfine splittings are of the order of an electronvolt in energy, about one hundred million times smaller than the effect in quark clusters. Most of this is due to the fact that atoms are between one and ten million times larger than quark clusters which implies that the energy splittings should be one to ten million times smaller on dimensional grounds. That they are yet smaller by a factor of ten to a hundred is because the electromagnetic force in atoms is intrinsically that much weaker than the quark forces in hadrons.

When discussing the intrinsic strength of the inter-quark forces and comparing QCD with QED it is important to specify the distances involved over which the colour force has to be transmitted. Although QED and QCD are mathematically almost identical at the outset, the replacement of *one* charge by *three* colours causes the two forces to spread out spatially in totally different ways (see Box 7.7). The theory implies that when coloured objects like quarks are less than a fermi apart the forces between them are almost nonexistent, and, in contrast, grow in strength as the quarks are pulled apart from one another (this is the basis of the scenario outlined on p. 100). The original SLAC experiments (p. 83) discovered this very phenomenon—high energy electrons scattered from quarks that were apparently free and yet stayed confined in clusters (Box 7.8). It is conjectured, but still not proved, that QCD predicts that the forces shoot up to infinite strength when quarks are separated by over a fermi. This infinite restoring force confines the quarks permanently in colourless clusters. Protons, pions, and related particles all have sizes of the order of 1 fermi, which is believed to be a consequence of this behaviour. I stress again that the complete details remain to be understood.

Box 7.7 QED and QCD: Very similar theories but far-reaching differences between the forces

QED

Charge emits photon.   Charge absorbs photon.

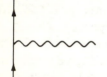

Photons can only be absorbed by another charge.

QCD

Colour charge emits gluon.   Colour charge absorbs gluon.

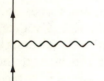

Gluon can only be absorbed by another colour charge.

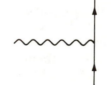

So far QED and QCD are similar.

The difference

In QCD the gluons also carry colour charge. This gives a new possibility:

Colour charge emits gluon.   Colour charge absorbs gluon.

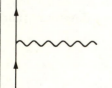

Gluon can only be absorbed by another colour charge.

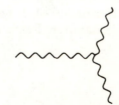

Box 7.7—*continued*

So the electromagnetic force between electrons

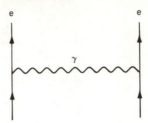

and the colour force between quarks

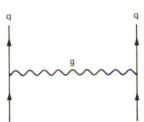

are different because a gluon can split into two gluons on the journey (QCD),

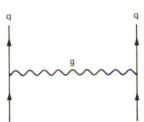

whereas a photon cannot split into two photons (QED) as they carry no electrical charge. It turns out that this causes the electromagnetic and interquark forces to behave quite differently. In particular the interquark force becomes weaker at short distances and strong at large distances, in agreement with data and explaining the paradox on page 84.

Box 7.8   How QCD helps to solve the quark force paradox

At high energy: free quarks in the proton but no escape

(Nearly free)
quarks in proton.

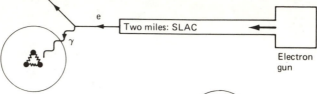

Quark recoils. The force holding it to the other
quarks grows and prevents it escaping.

The strong force isn't always strong

The picture of the proton changes with the energy of the photons that
see it.

1 GeV   10 GeV   100 GeV
(1950–65)   (SLAC 1968)   (Fermilab, CERN late 1970s)

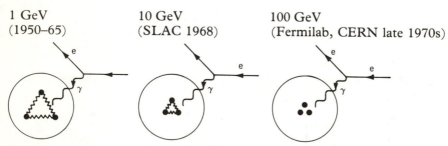

The photon sees quarks
tightly bound.
Genuine strong force,
complicated to deal
with mathematically.
1950–70, limited to low
energies: this tight
binding slowed up
theoretical progress.

At higher energies quarks appear almost free.
Force no longer strong.
QCD explains this and predicts that the force
gets weaker still as energy increases. 1979
experiments at CERN seem to confirm this
behaviour.

Large distance or low energy   QCD   Short distance or high energy

'Infra-red slavery'.
Interquark force strong, quarks
confined.

'Asymptotic freedom'.
Interquark force weaker (like
weak and electromagnetic in
strength at extremely high
energies and short distances).

8 The Weinberg–Salam model of the weak force

History

The most familiar manifestation of the weak force is the radioactive β-decay of nuclei. Such a transmutation was responsible for Becquerel's 1896 discovery of the β-particles (electrons) produced in the decay of uranium nuclei.

Nuclei are clusters of neutrons (n^0) and protons (p^+) and it is their neutrons that are the source of nuclear β-radioactivity:

$$n^0 \rightarrow p^+ + e^- + \bar{\nu}$$

(the e^-, electron, is a 'β-particle' and $\bar{\nu}$ an antineutrino, successfully predicted by Pauli in 1931 to explain the apparent imbalance of energy and momentum in such processes). By 1970 it had become apparent that neutrons and protons are in their turn clusters of quarks:

$$n^0(d^{-1/3}d^{-1/3}u^{+2/3})$$
$$p^+(u^{+2/3}d^{-1/3}u^{+2/3})$$

(the superscripts reminding us that up and down quarks have electrical charges that are fractions $\frac{2}{3}$ and $-\frac{1}{3}$ of a proton's charge. The combination ddu has total of zero as required in a neutron). The two clusters differ in that replacing one 'down' quark in the neutron cluster by an 'up' quark yields the cluster that we call a proton. The modern view is that the fundamental cause of β-radioactivity is the quark decay:

$$d^{-1/3} \rightarrow u^{+2/3} + e^- + \bar{\nu}$$

and so in the neutron

$$n^0(d^{-1\,3}d^{-1\,3}u^{+2/3}) \rightarrow p^+(u^{+2/3}d^{-1/3}u^{+2/3}) + e^- + \bar{\nu}$$

Elegant properties of the weak interaction appear in quark decay which are masked in neutron (let alone nuclear) decays. For example, the weak transmutation of a down to an up quark has identical properties to that of an electron into a neutrino and to that of a muon turning into another

105

Box 8.1   Quark weak interactions and neutron decay

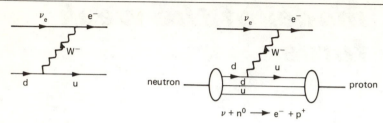

$$\nu + n^0 \longrightarrow e^- + p^+$$

A neutrino and a down quark interacting through the weak interaction produce an electron and an up quark. Bury the down quark in a neutron (two down quarks and one up quark) and this becomes the process

$$\nu + n^0 \to e^- + p^+$$

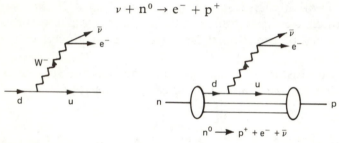

$$n^0 \longrightarrow p^+ + e^- + \bar{\nu}$$

Twist the figure around and we see the weak decay of a down quark:

$$d \to u + e^- + \bar{\nu}$$

Buried in the neutron this yields the weak decay of a neutron:

$$n^0 \to p^+ + e^- + \bar{\nu}$$

neutrino—called the 'muon–neutrino' to distinguish it from the former 'electron–neutrino'.

However in 1933 this was not known; neutron and proton were thought to be structureless building blocks of matter and it was in attempting to understand *neutron β-decay* that Fermi produced the first embryonic theory of the weak force. Although superceded in the last decade, his line of attack was essentially correct and is an instructive illustration of the progression from conception to birth of a fully fledged testable (correct?) theory of the weak force that combines it with electromagnetism.

Fermi's theory of neutron β-decay was inspired by quantum electro-dynamics where a neutron or proton absorbs a photon at a single point in space-time and preserve its electrical charge. Fermi proposed that something analogous occurred in β-decay: the change of charge in the decay of the neutron into a proton is caused by the emission of an electron and an antineutrino *at a point*.

The electron and antineutrino each have spin $\frac{1}{2}$ and so their combination can have spin 0 or 1. The photon by contrast has spin 1. By analogy with electromagnetism Fermi had (correctly) supposed that only the spin 1 combination emerged in the weak decay. To further the analogy in 1938 O. Klein suggested that a spin 1 particle ('W boson') mediated the decay—this boson playing a role in weak interactions like that of the photon in electromagnetism (Box 8.2).

In 1957 Julian Schwinger extended these ideas and attempted to build a unified model of weak and electromagnetic forces by taking the Klein model and exploiting an analogy between it and Yukawa's model of nuclear forces. As the π^+, π^-, and π^0 are exchanged between interacting particles in Yukawa's model of the nuclear force, so might the W^+, W^- and γ be in the weak and electromagnetic forces.

However, the analogy is not perfect in that the strong nuclear forces are charge-independent (π^+, π^-, and π^0 exchanges give comparable strengths) whereas the weak and electromagnetic forces manifestly are not: the forces mediated by W^+ and W^- appear to be more feeble than the electromagnetic force. Schwinger realised that this difference in strength is subtle and to some extent illusory. When we think of the weak force as feeble it is because we are describing it in the Fermi way where the interaction occurs at a single point in space and time. In the Klein and Schwinger picture the interaction occurs between distant particles much as the electromagnetic interaction does, the photon messenger being replaced by a W particle. The apparent strength of the weak force depends not only on the strength of the W interaction with matter but also on the mass of the W. The more massive the W so the more feeble would the interaction appear to be (see also Box 4.6).

If the W boson couples to the neutron–proton and to the electron–neutrino with electromagnetic strengths, then its mass must be of the order of 10–100 GeV if the neutron half-life and other observed low-energy interaction effects, such as the behaviour of neutrinos, are to be understood. The most recent developments in the theory uniting weak and electromagnetic interactions predict that these W bosons weigh about 82 GeV. Experiments capable of producing such massive objects recently began at CERN, and on 21 January 1983 their discovery was announced; their mass precisely as predicted.

The search and discovery of the W is described on p. 121. On a first reading you could profitably go their directly. The intervening pages give a more detailed description of some of the ideas in the electroweak theory and the experiments during the last decade that culminated in the discovery of the W boson.

Box 8.2   Weak decay

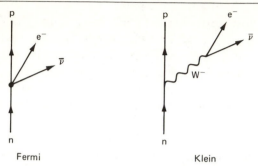

The Fermi and Klein models of β decay.

Box 8.3   The puzzle of the electron and proton electrical charge: episode 1

The electrical charges of electron and proton are opposite in sign and *exactly* equal in magnitude. This is the reason that atoms with the same number of electrons and protons have a net charge of zero, a fact that is so familiar that we don't give much thought to it. Yet this is really an astonishing phenomenon. Leptons and nuclear matter have appeared to be totally independent of each other in everything that we have discussed so far and yet the charged leptons and protons have precisely the same magnitudes of charge.

The simplest explanation is to suppose that electrical charge is some sort of external agent, a property of space perhaps, that is attached to matter in discrete amounts. Thus if we start with electrically neutral matter (neutron and neutrino), then the addition of the unit of charge to the neutron or removal from the neutrino will yield charged particles with exactly balanced amounts of charge. These are the proton and electron.

This was the sort of idea in Fermi's mind in 1933. The equality of charges is rationalised by relating the electron and proton to neutral partners, the neutrino and neutron respectively. Thus we see the first emergence of the idea of families; two leptons (neutrino and electron) and two particles of nuclear matter (neutron and proton).

The idea of families persists today but with down and up quark replacing the neutron and proton family. However, the idea of adding charge to fundamental neutral matter has been lost because the neutron is now known to be built from quarks which are themselves charged. The equality of electron and proton charges is therefore resurrected as a fundamental puzzle, suggesting that there exists a profound relationship between quarks and leptons.

A new form of weak interaction is predicted

In quantum electrodynamics a photon can be emitted or absorbed by an electrically charged particle. The chance that this happens is bigger the larger is the particle's charge. The charges of electron, proton, and uranium nucleus are proportional to the *numbers* -1, $+1$, $+92$. Refer to Box 8.3.

If neutrinos are fired at neutrons they convert into electrons and the neutron converts into a proton

$$\nu^0 + n^0 \rightarrow e^- + p^+$$

which is similar to the electromagnetic process

$$e^- + p^+ \rightarrow e^- + p^+$$

The Feynman diagrams illustrate how the photon and W boson play corresponding roles. The following weak interaction can occur instead of electromagnetic scattering in an electron–proton collision.

$$e^- + p^+ \rightarrow \nu^0 + n^0$$

in which case a W^- is exchanged.

Just as electron–photon coupling is described by a number (Box 4.6B), so could we denote the $v^0 \leftrightarrow e^- W^+$ and the $e^- \leftrightarrow v^0 W^-$ interactions by numbers. However that would be closing our eyes to an obvious further symmetry in nature: these processes are *not* independent of one another, in fact they have the same strength. Can we build this symmetry into our theory at the outset?

This can be done if we describe the electron and neutrino by a single entity—a matrix (see Box 8.4) with two entities:

Box 8.4   Matrices

There are many phenomena whose mathematical description necessitates more than just real numbers. One such generalisation of numbers is known as 'matrices'. These involve numbers arrayed in columns or rows with their own rules for addition and multiplication. Addition is no surprise:

$$\begin{pmatrix} a & b \\ c & d \end{pmatrix} + \begin{pmatrix} A & B \\ C & D \end{pmatrix} = \begin{pmatrix} a+A & b+B \\ c+C & d+D \end{pmatrix}$$

but multiplication is less obvious, it involves the product of all elements of intersecting rows and columns

$$\begin{pmatrix} a & b \\ c & d \end{pmatrix} \times \begin{pmatrix} A & B \\ C & D \end{pmatrix} = \begin{pmatrix} aA+bC & aB+bD \\ cA+dC & cB+dD \end{pmatrix}$$

The above matrices are 2×2—two rows and two columns—and from this generalised perspective, conventional numbers are 1×1 matrices! (or less trivially $N \times N$ matrices with the north-west to south-east diagonal elements all identical and all other entries zero).

Matrices can have any number of rows and columns and the number of rows need not equal the number of columns. Thus we can have column matrices such as

$$\begin{pmatrix} A \\ C \end{pmatrix} = A \begin{pmatrix} 1 \\ 0 \end{pmatrix} + C \begin{pmatrix} 0 \\ 1 \end{pmatrix}$$

When a 2×2 matrix multiplies such a column matrix the result is the same as above with B and D thrown away.

$$\begin{pmatrix} a & b \\ c & d \end{pmatrix} \times \begin{pmatrix} A \\ C \end{pmatrix} = \begin{pmatrix} aA+bC \\ cA+dC \end{pmatrix}$$

$$\left(\begin{array}{c} \text{chance of being neutrino} \\ \text{chance of being electron} \end{array}\right)$$

so that $\begin{pmatrix} 1 \\ 0 \end{pmatrix}$ represents a neutrino, $\begin{pmatrix} 0 \\ 1 \end{pmatrix}$ represents an electron. The W^+ and W^- are represented by the following 2×2 matrices:

$$W^+ = \begin{pmatrix} 0 & 1 \\ 0 & 0 \end{pmatrix} \qquad W^- = \begin{pmatrix} 0 & 0 \\ 1 & 0 \end{pmatrix}$$

The theory then requires that interactions of common strength among particles are described by multiplying the matrices together.

Using the multiplication rules shown in Box 8.4, we can first of all check that these matrices do indeed faithfully represent the pattern of interactions observed.

$$\begin{pmatrix} 0 & 1 \\ 0 & 0 \end{pmatrix} \times \begin{pmatrix} 0 \\ 1 \end{pmatrix} = \begin{pmatrix} 1 \\ 0 \end{pmatrix}$$

represents $\qquad W^+ + e^- = \nu^0 \quad \checkmark$

while

$$\begin{pmatrix} 0 & 0 \\ 1 & 0 \end{pmatrix} \times \begin{pmatrix} 1 \\ 0 \end{pmatrix} = \begin{pmatrix} 0 \\ 1 \end{pmatrix}$$

represents $\qquad W^- + \nu^0 = e^- \quad \checkmark$

These two processes have the same strength. Furthermore

$$\begin{pmatrix} 0 & 1 \\ 0 & 0 \end{pmatrix} \times \begin{pmatrix} 1 \\ 0 \end{pmatrix} = 0$$

implies that $W^+ + \nu^0$ does not happen. Nor do W^- and e^- interact together, in agreement with the matrix multiplication

$$\begin{pmatrix} 0 & 0 \\ 1 & 0 \end{pmatrix} \times \begin{pmatrix} 0 \\ 1 \end{pmatrix} = 0$$

These matrix multiplications are in one-to-one correspondence with the set of observed interactions and so appear to be the mathematics needed to describe the weak interaction. Indeed if one attempts to build a theory of the weak interaction by imitating the successful quantum electro-dynamic theory of electromagnetism one is inexorably led to the introduction of these matrices and that *all possible* multiplications represent physical processes. This is an important result: there are possible multiplications that we have not considered so far.

For example, let us see what happens if we multiply the W^- matrix by the W^+ one: the resulting matrix will correspond to a particle produced in a W^-W^+ collision.

$$\underset{W^+}{\begin{pmatrix} 0 & 1 \\ 0 & 0 \end{pmatrix}} \times \underset{W^-}{\begin{pmatrix} 0 & 0 \\ 1 & 0 \end{pmatrix}} = \underset{?}{\begin{pmatrix} 1 & 0 \\ 0 & 0 \end{pmatrix}}$$

This has produces a matrix that we have not met before and so implies that a new particle exists. This is a partner to the W^+ and W^- and has no charge (it was formed by a W^+ and W^- interacting). This is not a photon as can be seen by studying its interaction with a neutrino:

$$\underset{?}{\begin{pmatrix} 1 & 0 \\ 0 & 0 \end{pmatrix}} \times \underset{+ \ \nu^0}{\begin{pmatrix} 1 \\ 0 \end{pmatrix}} = \underset{\nu^0}{\begin{pmatrix} 1 \\ 0 \end{pmatrix}}$$

Photons do not interact with the electrically neutral neutrino, whereas the matrices show that this new particle does. Thus the Schwinger model was on the right track but the neutral partner of the W^+ and W^- is not simply the photon: the four matrices $\begin{pmatrix} 1 & 0 \\ 0 & 0 \end{pmatrix}, \begin{pmatrix} 0 & 1 \\ 0 & 0 \end{pmatrix}, \begin{pmatrix} 0 & 0 \\ 1 & 0 \end{pmatrix}, \begin{pmatrix} 0 & 0 \\ 0 & 1 \end{pmatrix}$ correspond to *four* particles Z^0, W^+, W^-, γ. Consequently, a new form of weak interaction is prediced when the Z^0 is involved. The old familiar processes are known as 'charged' weak interactions (alluding to the charged W^+ and W^-), the new process is called a weak 'neutral' interaction. The Z^0 can be exchanged between a neutrino and a proton and cause the previously unobserved process $\nu^0 + p \rightarrow \nu^0 + p$ to occur.

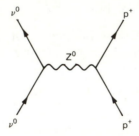

Thus neutrinos can interact with matter without swapping the charges around.

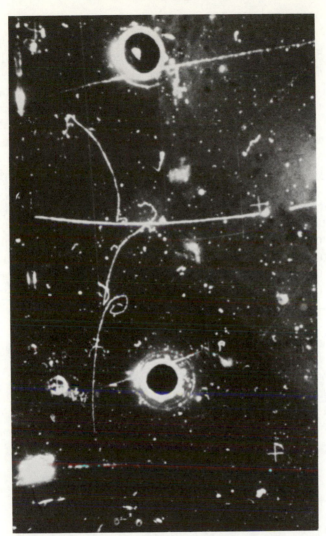

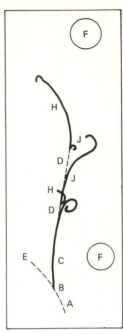

Photo 8A *Electrically neutral interaction of a neutrino* This undramatic-looking event played an important role in establishing the existence of the neutral current form of weak interaction that had been predicted by the Weinberg–Salam model of weak and electromagnetic interactions.

An antineutrino beam was fired into a chamber filled with liquid hydrogen. It leaves no track as it is electrically neutral. On hitting an electron in one of the hydrogen atoms the neutrino scattered leaving no track. The electron was kicked out of the atom (track C). This established the existence of the new interaction $\bar{\nu}_\mu + e^- \rightarrow \bar{\nu}_\mu + e^-$. The track is due to a recoiling electron as indicated by its radiating photons as it slows and spirals; the photons producing further electron and positron pairs. A schematic illustration is included, where: A = antineutrino, B = collision, C = scattered electron, D = high-energy photon, E = direction of scattered neutrino, F = lamps illuminating the chamber, H = positrons, J = electrons. (*Courtesy CERN*)

The discovery of neutral currents

In 'charged current' interactions, a neutrino converts into an electron which can be detected easily by virtue of its electrical charge. Electrically neutral particles, on the other hand, are notoriously difficult to detect—indeed we may recall that 25 years elapsed between Pauli proposing that neutrinos exist and Cowan and Reines detecting them. Furthermore they did so by observing the charged electrons that were produced as a result of the neutrinos interacting with matter. Small wonder that no one had ever seen processes where a neutrino comes in, scatters and remains a neutrino. However, if the model is correct then such 'neutral current interactions' must occur.

Experimentalists searched their data for examples of such a process. They looked for phenomena where a charged particle like an electron or proton suddenly moves off from rest when neutrinos are fired at them but no other visible phenomena accompany the interaction. By painstaking effort, evidence for the existence of such ephemeral interactions was obtained in 1973 at CERN by the Gargamelle collaboration, and has been confirmed by many subsequent experiments (Photos 8A, 8B).

Photo 8B   *Detecting neutrino interactions at CERN*   Neutrinos are very reluctant to interact with matter. To make them reveal themselves enormous detectors are required. This is the CDHS detector at CERN. Detectors today are larger than the early accelerators! (*Courtesy CERN*)

By calculating the Feynman diagrams for this process one can predict various properties of the interaction (how much energy the proton tends to absorb, what direction it tends to recoil in, whether it remains a proton or breaks up, and so on). One can also perform the experiment with electron targets instead of protons. All of these processes have been observed and extensively studied during the last ten years and all agree with the predictions of the theory if the Z^0, and W^+, W^- have masses of about 90 and 80 GeV respectively.

The discovery of these previously unknown forms of interaction was the first good evidence for the validity of the electroweak theory. By 1979 the quantitative successes of the theory, when confronted with the wide range of phenomena subsequently observed, led physicists to accept it as (almost) proved to be correct, even though the crucial production of W^+, W^-, and Z^0 bosons in the laboratory was still awaited.

Electromagnetic and weak interactions get together

I have glossed over some points of detail in the description of the weak interaction so far. The essential features have been correctly described but the following additional remarks are needed to show how the weak and electromagnetic interactions are wedded.

Recall that it was the fact that the weak interaction caused $\nu_e \leftrightarrow e$; $\nu_\mu \leftrightarrow \mu$ for leptons, and $u \leftrightarrow d$ for quarks that led to the introduction of matrices. Specifically, the electron and its neutrino (or the muon and its neutrino or the up and down quarks) are supposed to form doublets of 'weak isospin'

$$Q \equiv \begin{pmatrix} u \\ d \end{pmatrix}, \quad L_1 \equiv \begin{pmatrix} \nu_e \\ e^- \end{pmatrix}, \quad L_2 \equiv \begin{pmatrix} \nu_\mu \\ \mu^- \end{pmatrix}$$

In 1953 Gell-Mann and Nishijima had noted, in a different context, that the electric charges of particles in isospin doublets are in general given by:

$$\text{charge} = \pm \frac{1}{2} + \frac{Y}{2}$$

where $\pm \frac{1}{2}$ is for the upper or lower members of the pair while Y is an additive quantum number called 'hypercharge'. Leptons have different electrical charges from quarks and this is a result of the lepton doublets $L_{1,2}$ having $Y = -1$ (hence charge $= 0$ or -1 as observed), whereas the quark doublet has $Y = \frac{1}{3}$ (hence charges $\frac{2}{3}$ and $-\frac{1}{3}$). The weak interaction model described so far has taken no account of this hypercharge degree of freedom—that quarks and leptons have different values of it. Sheldon Glashow in 1961 was the first to do so, and produced the first viable

---

Box 8.5   SU(2)

The astute reader may have noticed that the collision of a W^+ and W^- could be represented either by:

$$W^+ \qquad\qquad + W^-$$

$$\begin{pmatrix} 0 & 1 \\ 0 & 0 \end{pmatrix} \times \begin{pmatrix} 0 & 0 \\ 1 & 0 \end{pmatrix} = \begin{pmatrix} 1 & 0 \\ 0 & 0 \end{pmatrix}$$

or by

$$W^- \qquad\qquad + W^+$$

$$\begin{pmatrix} 0 & 0 \\ 1 & 0 \end{pmatrix} \times \begin{pmatrix} 0 & 1 \\ 0 & 0 \end{pmatrix} = \begin{pmatrix} 0 & 0 \\ 0 & 1 \end{pmatrix}$$

Which of these represents the particle produced in the collision?

Deeper aspects of the theory than we can go into here, require that the 2×2 matrices must be restricted to those whose top left to bottom right diagonal numbers sum up to zero ('traceless matrices'). These are:

$$W^+ = \begin{pmatrix} 0 & 1 \\ 0 & 0 \end{pmatrix}; \quad W^- = \begin{pmatrix} 0 & 0 \\ 1 & 0 \end{pmatrix}; \quad W^0 = \begin{pmatrix} 1 & 0 \\ 0 & -1 \end{pmatrix}$$

and so W^0 is represented by the difference of the two possibilities above. These 2×2 traceless matrices are known as SU(2) matrices in mathematical jargon. The analogous $N \times N$ traceless matrices are called SU(N) matrices. Just as SU(2) contains $3 = 2^2 - 1$ so will SU(3) have $8 = 3^2 - 1$ (see Chapter 10).

The sum of the two matrices is $\begin{pmatrix} 1 & 0 \\ 0 & 1 \end{pmatrix}$ and is known as a U(1)

matrix. The fact that W^+ and W^- can produce a photon in addition to the W^0, is related to the two independent matrices arising from a W^+W^- collision. The total theory of weak (W) and electromagnetic (γ) interactions is mathematically an SU(2) $\times$ U(1) theory.

---

model that brought the weak and electromagnetic interactions together (for which he shared the Nobel Prize in 1979).

The matrix mathematics that described the weak isospin is known as 'SU(2)' (the 2 refers to the doublet nature of the Q, L_1, L_2 above, or of the 2×2 representations of the W particles. SU is a mathematical classification, see Box 8.5). Thus the matrix theory of the weak interaction described so far is called an SU(2) theory.

The hypercharge on the other hand is a real number. In matrix jargon we think of it as a 1×1 matrix and so the mathematics involving such numbers is called U(1)—analogue of the SU(2). Combining the weak isospin and hypercharge yields an SU(2) $\times$ U(1) theory.

The new feature that enters, as compared with the previous treatment, is that in addition to the W^+, W^-, W^0 of the SU(2) piece (which couple to matter with a common strength g_2) there is a fourth particle from the U(1) piece, an electrically neutral 'B^0' (which couples to matter with a strength g_1). The relative strengths g_1 and g_2 are quite arbitrary and are conventionally described in terms of a parameter θ_W (known paradoxically as the Weinberg angle even though this model was first constructed by Glashow in 1961).

Now we are almost home and dry. The physical photon couples to electric charge and is a quantum superposition of W^0 and B^0. The W^+ and W^- transport the 'charged' weak interaction known for 50 years. In addition there is a new 'neutral' weak interaction carried by Z^0—the orthogonal superposition of W^0 and B^0:

$$
\mathrm{SU(2)} \begin{cases} W^+ \\ W^- \\ W^0 \end{cases} \quad \begin{rcases} W^+ \\ W^- \end{rcases} \text{charged weak interaction}
$$

$$
\begin{matrix} \times \\ \mathrm{U(1)} \quad B^0 \end{matrix} \Bigg\} \quad \text{mix} \begin{cases} \gamma & \text{electromagnetism} \\ Z^0 & \text{neutral weak interaction (predicted)} \end{cases}
$$

The way that this mixing occurs is beyond the scope of the present text. It is as if the W^0 and B^0 are represented by two axes in some abstract space. The γ and Z^0 are represented by two orthogonal vectors at some angle θ_W relative to these axes (Box 8.6). This is the same θ_W that is related to the relative strengths g_1 and g_2.

Box 8.6   The Weinberg angle

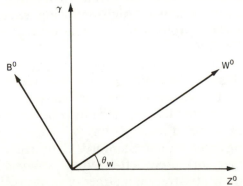

The physical γ and Z^0 are mixtures of the B^0 and W^0. If $\theta_W = 0$ then $W^0 \equiv Z^0$ and $\gamma \equiv B^0$; the weak and electromagnetic interactions would not be mixed together. Empirically $\sin^2 \theta_W \simeq \frac{1}{5}$ and the photon is a superposition of both B^0 and W^0.

Does electron scattering respect mirror symmetry?

Not only can the Z^0 be exchanged by neutrinos interacting with matter, but it can also be exchanged by electrons. This can give rise to the following interaction:

$$e^- + p^+ \rightarrow e^- + p^+:$$

(a)

This is going to be swamped by the much more probable contribution from photon exchange

$$e^- + p^+ \rightarrow e^- + p^+:$$

(b)

and so had never been seen in electron–proton scattering experiments. With the development of the electroweak theory, interest grew in the possibility of performing a very high-precision experiment to study electron scattering from protons to see if any evidence for this extra contribution could be found.

Charged weak interactions do not respect mirror symmetry (Box 4.7). The Weinberg–Salam model uniting weak and electromagnetic interactions required also that there is no mirror symmetry in neutral current interactions, specifically that left-handed electrons interact with Z^0 more readily than do right-handed ones.

The two-mile long accelerator at Stanford, California, can produce high energy electrons that corkscrew left-handed (as a neutrino) or right-handed. Now fire these electrons at a target of protons. It was found that the left-handed electrons have a greater tendency to interact with the protons than do the right-handed ones. The left-handed excess was only one event in every ten thousand or so, but this is a large effect within the sensitivity of the experiment.

We have here a new way of distinguishing the real world from the mirror world—there are both left-handed and right-handed electrons but it is the former that prefer to interact in the real world. A new piece of evidence for a left-handed preference in nature had been obtained. The significance of this discovery was that the model combining weak and electromagnetic interactions had correctly predicted *both* that the left-handed electrons should win *and* by how much.

It is this observation of the left-handed excess which provided the first evidence that the weak neutral current couples to electrons, and the first *direct* proof that it does not respect mirror symmetry.

Do atoms respect mirror symmetry?

The archetypical example of the electromagnetic interaction is the formation of atoms from electrons surrounding nuclear protons, held by the electromagnetic attraction of opposite charges. In the combined weak and electromagnetic interactions there will be an additional force between the electrons and protons in atoms, namely the 'weak neutral force' due to their mutual exchange of the Z^0 particles. At low energies this force is very much weaker than the prominent electromagnetic force and so detection of its effects in atomic phenomena requires very high precision experiments.

When an electron jumps from a high-energy atomic orbital to another of lower energy, the excess energy is radiated off as light. If the atom is in a magnetic field we can measure the direction that the individual photons are emitted relative to the field. If mirror symmetry is respected, then the probability that a photon is emitted in a certain direction can be calculated. What in effect is being looked for in the experiment is evidence for a small deviation from the expected distribution.

These experiments are extremely difficult to perform at the level of precision required if subtle effects are to be detected. There has been controversy over the results for some years, as different groups of experimenters obtained results that appeared to be at odds with one another. However, it seems that the dust might now be settling and it is just possible that an effect *is* being seen as predicted by the theory. It is interesting to reflect how a theory whose development was motivated by phenomena at extremely high energies turns out to have consequences in low-energy atomic physics.

This is a nice example of how the secrets revealed in one sort of experiment are not just applicable to that type of experiment or field of research. Instead, a true law of nature, once discovered, is applicable anywhere and for ever.

Higgs bosons

The model of weak and electromagnetic interactions that Glashow set up is a matrix generalisation of standard electromagnetic theory. The relativistic quantum theory of the latter, Quantum Electrodynamics, had been developed several years earlier and implied that photons are massless. This is of course quite satisfactory because photons do indeed have this property, but is a problem if you try to imitate it in building a theory of the weak force: Glashow's model seemed to imply that photons *and* the W^+, W^-, and Z^0 were all massless. This would be a disaster because even in the 1960s there was good evidence that the W and Z particles could not be light (a light W or Z would have been produced copiously in interactions much as photons are produced). So Glashow's model had no explanation of why the W and Z had mass whereas the photon was massless.

Stimulated by some developments in solid state physics and quite independent of the above attempts to build theories of the weak interaction, in 1964 Peter Higgs in Scotland and independently, Robert Brout and Francois Englert in Belgium, discovered that there was a loophole in the line of theoretical argument that required the photon to be massless. Higgs showed that it could have a mass if there also exist in nature massive spinless particles (known as Higgs bosons after their inventor).

The discovery of the 'Higgs mechanism' might seem to be rather academic because nature does not appear to exploit this loophole: empirically, photons are massless. But then Abdus Salam and Steven Weinberg independently suggested that nature might exploit this loophole for weak interactions. They showed that the photon stayed massless while the W and Z gained quantifiable masses.

However there was one remaining problem. While this created a theory that is satisfactory at low energies, no one was quite sure whether or not it made any sense when applied to high energy interactions—it seemed to predict that certain processes would occur with infinite probability!

The final link was provided by a young Dutchman, Gerhard 't Hooft in 1971. He proved that the theory *is* self-consistent and gives finite probabilities for physical processes at all energies.

If Higgs' mechanism is the way that W and Z particles become massive then spinless Higgs particles ought to exist. The properties of these particles are in dispute among the theorists at present. No one is able to predict their masses with any degree of confidence and there is even one school that argues that such particles are not elementary but are composed of yet more fundamental particles. Further understanding in this area will probably require prior experimental discoveries to be made.

Discovery of the W boson

At the risk of boring readers who ploughed through the last 13 pages, I shall first make a brief summary in case you have jumped here directly.

Schwinger's unified model involving W^+, W^- and γ had promising features but also had some flaws. Most obvious of the problems was that W^+ and W^- both interact with neutrinos but photons do not, or at best only indirectly. A further neutral boson was required ('Z^0') which can directly couple to neutrinos as well as to other matter. This idea was first developed by Sheldon Glashow in 1961 and then extended by Steven Weinberg and Abdus Salam in 1967. They showed that the masses of Z^0, W^+ and W^- could be mutually related, in particular the Z^0 cannot be lighter than the W^+ and W^- (and so *cannot* be the photon).

The electrically neutral weak interaction mediated by the Z^0 was discovered in 1973 and a subsequent series of tests led to the following predictions. If weak interactions are really intrinsically united with electromagnetism in this way then

$$m_{W^+} = m_{W^-} = 82 \pm 2 \text{ GeV}$$

$$m_{Z^0} = 92 \pm 2 \text{ GeV}.$$

For their formulation of the theory, Glashow, Salam, and Weinberg shared the 1979 Nobel Price for physics, even though neither W or Z had been detected! Their theory had, after all, predicted a new phenomenon—the neutral weak interaction—which was, of itself, a major step.

The task now was to produce the W and Z particles which are the quanta of weak radiation and bundles of weak energy. No machine existed powerful enough to do this.

CERN had just built a super proton synchrotron (SPS) capable of accelerating protons to energies of several hundred GeV. Then came the idea of creating antiprotons in a laboratory next to the SPS, accumulating them until huge quantities were available, and then injecting them into the SPS. The SPS is a circle of about 4 miles circumference: protons swing around in one direction and antiprotons in the other. When they meet head on a cataclysmic annihilation of matter and antimatter can occur and for a fraction of a second conditions are similar to those present in the Big Bang. At these extremes, the W and Z particles are expected to appear.

The major difficulty was in storing the antiprotons. These are examples of antimatter, and as the machine, laboratory, physicists, and apparatus are all made of matter it was essential to keep the antiprotons away from them and in a total vacuum. The enthusiasm of Carlo Rubbia, whose idea it was, and the brilliance of the machine designer, Simon van der Meer, combined to make the dream come true.

By the autumn of 1982 it was at last possible to smash the protons and antiprotons against each other. Usually they fragment, their quarks and antiquarks creating showers of particles—one shower following the line of the proton, the other of the antiproton. But occasionally one quark and one antiquark can mutually annihilate and be converted into radiant energy. If weak and electromagnetic interactions are united at these extreme energies, then bundles of energy will be manifested as W or Z bosons no less frequently than as photons. The W and Z are highly unstable and within 10^{-23} seconds, decay into electrons, positrons or neutrinos which can shoot out perpendicular to the direction of the incident protons. Out of more than a million collisions, nine precious examples of such events had been found by January 1983.

The characteristics of these nine events are as conclusive proof as a partial fingerprint can be at the scene of a crime. Their abundance suggests that the W is produced as easily as photons, and its mass appears to be consistent with the theoretical expectations.

It begins to look as if the intuition developed during the 1970s is correct. The electromagnetic force and the weak forces appear to have more to do with each other than either does with gravity or the strong nuclear force. In the coming months and years there will be much detailed study of these new phenomena in order to determine just how similar or otherwise they are. This discovery has given further impetus to the belief that at even more extreme energies the strong nuclear force also will unite with the electroweak force.

Conclusions

In this chapter we have witnessed one of the great accomplishments of recent years. The formulation of a successful theory of weak interactions is in its own right a major feat, but the fact that it is so similar to, and subsumes, quantum electrodynamics is far more exciting.

In Chapter 7 we met the discovery that any 'flavour' of quark could occur in three 'colours'. Just as the weak interaction theory emerged when we studied flavour *doublets* (i.e. electron and neutrino are two lepton flavours) one is tempted to ask what would happen if we played an analogous game with colour *triplets*, for example

$$\begin{pmatrix} \text{red quark} \\ \text{yellow quark} \\ \text{blue quark} \end{pmatrix}$$

and built an SU(3) theory involving 3×3 matrices.

The result is a theory similar to quantum electrodynamics but based on three colours. In place of one photon or three weak force carriers (W^+, W^-, Z^0) we now find eight colour carriers ('gluons'). This theory is quantum chromodynamics whose properties were described in Chapter 7. Theories generated this way are called gauge theories. Electromagnetic and weak interactions are described by U(1) and SU(2) gauge theories; quantum chromodynamics is an SU(3) gauge theory.

Why (or if) nature exploits the Higgs mechanism in the SU(2) case but not in the others is an open question. Apart from this asymmetry the theories are very similar mathematically. Physicists suspect that this similarity is too marked to be an accident and believe that all of these forces are intimately related in some way. This had recently led to the development of Grand Unified Theories (GUTS for short) of all the forces. Before describing these theories we should bring ourselves up to date with some startling discoveries during the last decade of new varieties of subnuclear matter, which played a seminal role both in developing the electroweak theory and in turn stimulating the construction of Grand Unified Theories.

Postscript January 1984:
The W has now been seen more than 50 times. The Z was discovered in June 1983 and a dozen examples have been recorded subsequently. Their masses and general behaviour are in complete accord with the theoretical predictions.

9 Charm and new discoveries

The theory of electromagnetic and weak interactions described in the previous chapter was built upon the observation that leptons and quarks form pairs

$$Q_1 \equiv \begin{pmatrix} u \\ d \end{pmatrix} \quad L_1 \equiv \begin{pmatrix} \nu_e \\ e^- \end{pmatrix}$$

$$L_2 \equiv \begin{pmatrix} \nu_\mu \\ \mu^- \end{pmatrix}$$

This is fine for the leptons and the up and down quarks, but leaves the strange quark in isolation and out of the weak interaction. In his original paper postulating quarks, Gell-Mann alluded to the possibility of a fourth quark (Box 9.1) that formed a pair with the strange quark, 'by analogy with the leptons'. This idea was briefly pursued by Glashow, among others, but was then dropped because not a single hadron containing a 'charmed' quark was found.

Although the idea was not pursued, it was not entirely forgotten. Glashow had been at Caltech when Gell-Mann was developing his quark ideas, and it was during that period that Glashow had himself been developing his theory of weak interactions in close consultation with Gell-Mann. His theory was initially built for leptons where it worked perfectly, successfully predicting the existence of neutral weak interactions. But when applied to quarks it failed—in addition to the desired neutral currents it also predicted that another, unwanted, variety occurred. These unwanted interactions involved not the satisfactory $\nu d \to \nu d$ but required that down and strange quarks could transmute into one another, $\nu d \to \nu s$. Such 'strangeness-changing neutral interactions' do not occur in nature.

Glashow mused that everything works fine for the leptons: the magic ingredient being that the leptons form *two* pairs. He revived the idea of the fourth quark and with John Iliopoulos and Luciano Maiani showed that everything with the weak interaction theory would be perfect for

Box 9.1 **Leptons and quarks in weak interactions: the charmed quark**

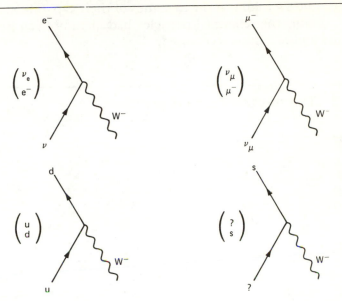

The properties of quark weak interactions are identical to those of leptons. As far as weak interactions are concerned the up–down quark doublet and the electron–neutrino lepton doublet are identical. If there exists a fourth quark ('charmed quark') at '?', the parallel would be very striking.

both leptons *and* quarks if there exists a charmed quark with charge $\frac{2}{3}$, like the up quark, and if the quarks form two pairs similar to the leptons (Box 9.2):

$$Q_1 \equiv \begin{pmatrix} u \\ d \end{pmatrix} \qquad L_1 \equiv \begin{pmatrix} \nu_e \\ e^- \end{pmatrix}$$

$$Q_2 \equiv \begin{pmatrix} c \\ s \end{pmatrix} \qquad L_2 \equiv \begin{pmatrix} \nu_\mu \\ \mu^- \end{pmatrix}$$

The unavoidable consequence of this is that there exist scores of hadrons containing charmed quarks—so why had no one ever found any? If the charmed quark is very massive, then hadrons containing one or more charmed quarks or antiquarks could be so heavy that they could not have been produced in existing accelerators. However, it was touch and go.

For the GIM theory (pronounced 'Jim'), as it is known, to work, the charmed quark had to be heavy but not *too* heavy. It seemed possible that charmed particles could possibly be light enough to be produceable by existing machines working at the extremes of their capabilities. Indeed it was even possible that charmed particles had already been produced in experiments and not been recognised.

Box 9.2   The neutral current problem

Leptons

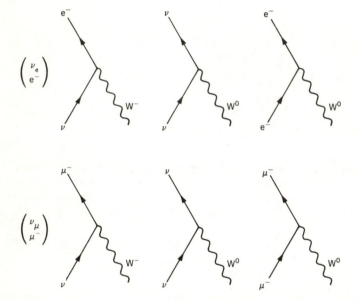

The leptons seem to exist in two separate pairs. No mixing occurs between the two pairs, i.e. transitions like

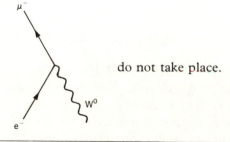

do not take place.

Box 9.2—*continued*

Quarks

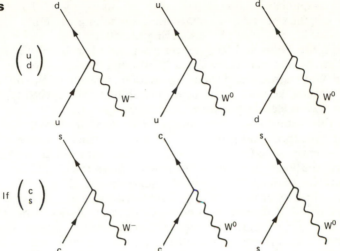

The up and down quarks form a pair. If a charmed quark existed then a second pair would arise. Transitions like this:

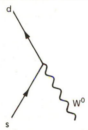

There is good evidence that the transition $d \leftrightarrow sW^0$, does not occur in nature.

Such a transition is called a 'strangeness-changing neutral current': 'neutral' because the down and strange quarks have the same electrical charge, and 'strangeness-changing' because they have different strangenesses. Such transitions are avoided if there are two 'generations' of quarks in analogous fashion to the way that two 'generations' of leptons do not admit the transition

would not occur: the quarks and leptons would behave very similarly.

These pairings are now called 'generations' of quarks and leptons. (u, d) or (e^-, ν) form the first generation. (c, s) or (μ^-, ν) form the second generation.

(a)

(b)

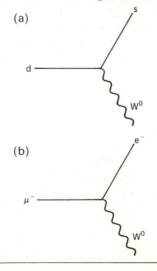

Box 9.3 Annihilating matter and antimatter

One of the significant features that led to the discovery of charmed particles was the change of emphasis that was taking place in experimental high-energy physics during the 1970s.

Through the 1950s and '60s the central puzzle had been why so many strongly interacting particles existed. With the recognition of the Eightfold Way patterns and their explanation in terms of the quark clusters it was of central importance to establish if the spectroscopy of the states was indeed of this form.

To do this, protons were fired at stationary targets of matter (essentially neutrons and protons). The debris consisted mainly of strongly interacting particles. The experiments were designed to study this debris and look for examples of new particles that could be compared with the expectations of the quark model. The debris consisted primarily of particles that could be easily produced from protons, mainly things built from up and down flavours of quarks. Strange quarks were light enough that pairs of strange quarks and antiquarks could be produced from the available energy in the collision and so strange mesons or baryons could also be produced and studied. Charmed quarks are very massive. As the everyday world is built from up and down flavours, a charmed quark and charmed antiquark have to be produced together and this is very difficult; hence the failure to find charmed particles in these experiments.

Then in the late 1960s attention was focussed on experiments involving leptons. One of the first of these was the classic experiment at Stanford where electron–proton collisions showed the first clear view of the quarks inside the proton (p. 83).

The most significant advances came with the development of 'storage rings' where electrons and their antiparticles, positrons, were stored and then collided head-on. Most frequently they scattered from one another, but occasionally the electron and positron, being matter and antimatter, annihilated each other and produced pure energy. When enough energy is concentrated in a small region of space, matter and antimatter can be produced, in particular a quark and an antiquark.

The early machines of this sort were at Orsay near Paris and at Frascati near Rome. The maximum energy that the Orsay machine possessed was just sufficient to produce a strange quark and strange antiquark together. It was found that the chance of doing this increased dramatically when the total energy in the collision was just over 1000 MeV (1 GeV). This was because the ϕ meson, which is a bound state of the strange quark and the strange antiquark, has this mass, and was being produced by the electron–positron annihilation.

At Frascati the experimentalists were able to study electrons and

Box 9.3—*continued*

positrons annihilating each other up to a total energy of 3000 MeV (3 GeV). We must count them as extremely unlucky. As we now know, the mass of the ψ (the spin one bound state of a charmed quark and charmed antiquark) is very slightly greater than this (3095 MeV to be precise) and so they did not have enough energy in their machine to produce the beast. If someone in 1970 had created a theory that had convincingly predicted the ψ at that mass, then the Frascati physicists could have designed their machine slightly differently and managed to produce it. But, of course, no one had that information then. (In fact when the ψ was eventually discovered in 1974 with a mass slightly above the Frascati machine's maximum, the physicists at Frascati managed to squeeze every bit of available energy out of the machine and within a couple of days had seen the ψ themselves.)

Photo 9A *Electrons annihilate positrons at SPEAR* Electrons and positrons created at SLAC (photo 6A) can be brought together in the circular rings SPEAR. They meet head-on and annihilate. (*Courtesy SLAC*)

Box 9.3—*continued*

Photo 9B *Detecting the results of electron–positron annihilation at SPEAR* After the electron and positron have annihilated, new matter and antimatter can be created. This is detected by the array of electronic devices shown. Electrons enter at one end and positrons at the other end, the annihilation occurring in the tunnel behind Roy Schwitters. In this way the J/Ψ and charmed particles were discovered. (*Courtesy SLAC*)

The best place to look seemed to be in electron-positron annihilation experiments (Box 9.3). If this annihilation occurred with sufficient energy then a charmed quark and a charmed antiquark could be produced. Sometimes these might bind together to produce a meson built from a charmed quark and a charmed antiquark. During 1973 and 1974 some peculiar phenomena were seen in the behaviour of electron–positron annihilations at energies between 3 and 4 GeV. With hindsight it is astonishing that no theorist gained a share of a Nobel Prize by suggesting that here, staring us in the face, was the evidence for charm.

As with the original quark idea we have here another example of how scientific progress is not inexorably forwards and in unison. Apart from a few committed aficianados, charm was not where most physicists' attention was directed. Everything changed dramatically on 10 November 1974 with the discovery of the first example of a particle built from a charmed quark and a charmed antiquark—the J/ψ meson—nowadays referred to simply as the ψ. The significance of that discovery was such that physicists now rank that day alongside the annus mirabile 1932 (when the positron and neutron were both discovered). Not only was there scientific revolution but there were great personal dramas too.

The J/ψ meson

The first machine with sufficient energy available for electron and positron to annihilate and produce the ψ meson was the storage ring 'SPEAR' at Stanford, California (Photo 9A). (The electrons and positrons for this machine were provided by the two mile long electron accelerator in Photo 6A.) It was here that a team led by Burton Richter discovered the ψ meson during the weekend of 9–10 November 1974. On Monday 11 November the discovery was announced at a special meeting at Stanford. Attending this meeting was Prof. Samuel Ting from MIT. By a strange quirk of fate, Ting had already discovered this meson in an experiment that he had been performing at Brookhaven Laboratory, New York and had not yet announced it.

Ting had been leading a team studying collisions of protons and nuclei. They were not interested in the debris of pions that provided over 99% of the events; instead they were seeking those much rarer events where an electron and a positron were produced among the debris. By measuring the energies of the pair they could look to see if the electron and positron preferred to come out more frequently with some particular value for their energy sum than other values. If they did, then this would show that a meson had been produced in the collision which had subsequently decayed into the electron–positron pair.

Photo 9C *Samuel Ting and the J/Ψ discoverers at Brookhaven* The combined effort of the Massachusetts Institute of Technology and Brookhaven National Laboratory brought about the physics discovery of the decade made at Brookhaven National Laboratory—the discovery of the J/Ψ particle. The discoverers are seen with a graphic representation of the event. (*Courtesy Brookhaven National Laboratory*)

Photo 9D *The J/Ψ discoverers at SPEAR, Stanford* Gerson Goldhaber, Martin Perl, and Burton Richter, leaders of the combined Stanford and Berkeley collaboration that discovered the J/Ψ particle in November 1974. Professor Richter shared the Nobel prize for physics with Professor Ting (photo 9C) for his part in this joint discovery. (*Courtesy SLAC*)

In a sense, this was Richter's experiment in reverse. Richter sent the electron and positron in at the start and produced the ψ meson; Ting produced the meson and detected it by its subsequent decay into an electron and positron. This is summarized in Box 9.4.

This is a very difficult way to look for mesons. Like all high-energy physics experiments you have to look at an enormous number of events and perform a statistical analysis on them to see if a peak in the mass histogram is significant or not. You plot a histogram and find a few extra events in one particular region. As you accumulate events you find this excess consistently appearing. Gradually the small excess builds up into a large excess and then you feel that you might have found something.

Paradoxically, Ting had been gathering data during the summer of 1974 and found a *huge* enhancement in one small region of the histogram. It was too good to be true. If correct, he had discovered a meson with unexpected stability and which was substantially heavier than anything ever seen before (its mass of 3095 MeV is more than three times that of a

Box 9.4 J or Ψ meson production: the complementary discoveries of Richter and Ting

(a)

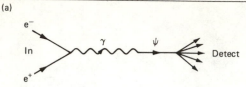

(a) Richter at SLAC produced it by annihilating electrons and positrons. He then detected its decay into hadrons.

(b)

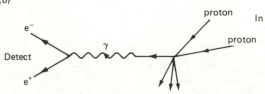

(b) Ting at Brookhaven produced it by using hadrons and detected its decay into electron and positron.

proton). The peak in Ting's distribution was so stunning that the first suspicion was that some quirk of the apparatus was artificially producing the enhancement rather than it being a genuine discovery.

During the autumn Ting's group were checking and rechecking their data. So with this intensive work going on back on the east coast of the USA, Ting sat in a lecture room at Stanford in California and heard that a collaboration of physicists from Berkeley and Stanford had just discovered a meson with precisely the mass and stability that his own data had suggested. This confirmed that Ting had discovered a genuine effect. He named it the J-meson and rapidly telephoned his group. The Stanford group named it the ψ meson (it later became known as the J/ψ meson).

The news travelled rapidly around the world and physicists at Frascati in Italy were able to confirm its existence within just a couple of days. Once found, the meson was so visible that it was soon being studied in laboratories the world over.

About ten days later, the team at Stanford had discovered an even more massive particle, the ψ' of mass 3684 MeV (nearly as massive as a helium nucleus!). This is formed from the charmed quark and antiquark in an excited state with total spin of one. During the following year a whole spectroscopy of states built from a charmed quark orbiting around a charmed antiquark were discovered at Stanford. These are called states of 'charmonium'.

Just as hydrogen exists in excited levels, and states of higher energy can decay into states of lower energy by emitting photons, so can the charmonium states of high mass (energy) decay into lower mass charmonium states by emitting photons. The only difference is one of scale: the photon emitted in the atomic case has an energy of a few electronvolts (eV) whereas in the charmonium case the photon energy is over a million times larger, being of the order of a hundred MeV.

Charmed mesons

The spectroscopy of mesons built from a charmed quark and charmed antiquark was uncovered during 1975. The lightest of these (the J/ψ) has a mass of 3095 MeV and the charmed quark is therefore of about 1500 MeV in mass (Box 9.5).

To complete the picture one wanted to find evidence for mesons built from a charmed quark (or antiquark) accompanied by one of the 'old-fashioned' up, down, or strange flavours. These so-called charmed mesons would be expected to be about 1850 MeV in mass (a charmed quark being 1500 MeV and an up or down quark being about 350 MeV). The J/ψ is too light to produce a pair of these mesons in its decay. (To do this would require at least two times 1850 MeV.)

It was then found that when an electron and positron collide with energy greater than about 4 GeV (4000 MeV) the probability of annihilation increased slightly, suggesting that charmed mesons were emerging in pairs. According to the theory that had predicted the existence of charm, the charmed quark was a partner of the strange quark in the weak interactions which implied that when a charmed meson decayed it should produce a strange meson in its decay products (Box 9.6) (for example a K meson).

In fact it was predicted that a charmed meson could decay into one strange meson accompanied by a pion. So the experimentalists looked at the debris to see if there was any hint of a K and a π meson being produced, such that if you lumped their energies together they preferred to centre around some particular value.

Unlike the J/ψ meson which had been startlingly obvious, charmed mesons were rather difficult to detect and no hint of them was found for over a year. It was not until the summer of 1976 that the data were clear enough to show that the charmed mesons had been found. Their masses were about 1875 MeV where the quarks were spinning antiparallel (analogues of the K) and about 2000 MeV for the parallel spin configuration (analogue of the K$\star$). Note again the systematic phenomenon where the parallel spin configuration is heavier than the antiparallel, compare page 101.

Box 9.5   J/Ψ and charmed particles: history

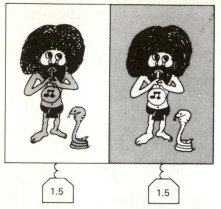

The J/ψ meson discovered in November 1974 was the first example of a particle containing a charmed quark.
It is built from a charmed quark and a charmed antiquark.
These each weigh about 1.5 GeV—50% more than a whole proton!—and give the J/ψ a mass of 3 GeV.

In 1976 the first examples of particles carrying manifest charm were seen.
A charmed quark and an up antiquark shown here yield a 'D-meson' mass 1.85 GeV.

Recently the charmed baryons have been isolated.
A charmed quark and two up quarks yields the Σ-charm baryon, mass 2.2 GeV (below).

Particles containing charm and strange quarks should also exist and are being looked for. The F-meson: charm and antistrange has been seen.

Box 9.6 Transmutation and charm

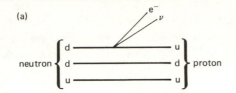

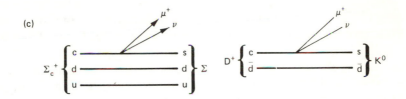

Radioactive transmutation of a neutron through beta decay was illustrated in Box 8.1. It occurs when a down quark (d) becomes an up quark (u) emitting an electron and a neutrino (technically an antineutrino).

A neutron is formed from two down quarks and one up quark, while a proton consists of one down and two up quarks. Consequently the act of converting a down quark into an up quark causes a neutron to become a proton (part a).

This act of transmutation involves all of the first generation of elementary particles, namely up and down quarks, electron and neutrino leptons (part bi). The analogous process involving second generation particles is shown in part (bii). Because the charmed quark is more massive than the strange quark, this second generation reaction tends to proceed in the opposite direction in nature, that is $c \rightarrow s + \mu^+ + \nu$. The μ^+ is the positively charged antiparticle of the muon. The direct analogue of the beta decay of the neutron is shown in part (ci). Experimenters have not yet detected this reaction although the decay of a charmed particle, a meson, formed from a quark (c) plus an antiquark ($\bar{d}$) has been seen (part cii).

So the masses were about right. Furthermore, the charmed meson decays into a strange meson precisely as predicted. This was a significant testimony to the existence of a charmed quark and also to it being the partner of the strange quark. This was a really major step; it showed that the theories developed over the previous years were indeed on the right track. It was an important clue in showing how the universe works.

The J/ψ significance

The discovery of the J/ψ was like finding the proverbial needle in a haystack. Once it had been found you knew exactly where to look for the predicted related charmed particles and one by one they began to emerge with all the properties that had been predicted for them. Thus, it was the J/ψ discovery that triggered the breakthrough in unravelling charm.

The charmed particles' existence had been predicted both on aesthetic and also scientific grounds. The aesthetic was that the leptons and the quarks now shared a common pattern—four of each, and forming pairs when they partake in the weak interactions. This already gives a tantalising glimpse that we may some day find that the leptons and quarks are profoundly related to one another (why else should they have such similarity?).

Furthermore, the charmed quark (and consequent particles containing it) had been required, in order that ideas uniting the weak and electromagnetic interactions could survive. The discovery that the charmed quark exists *and* with precisely the properties predicted in this theory gave impressive support that such an idea may have a substantial amount of truth in it.

The J/ψ discovery was significant in all of these respects. It is therefore no surprise to learn that the leaders of the two teams that discovered it (Richter at Stanford and Ting at MIT) were awarded the Nobel Prize for physics within only two years of the discovery.

Dramatic and significant as this was, there is one final feature of the J/ψ discovery that merits mention. This may turn out to be at least as significant as the above properties. The J/ψ meson is extremely stable. There were some predictions that such an object, if it existed, should be rather stable but when it was discovered, it turned out that it was *too* stable. It lived significantly longer than had been predicted by the theoreticians in advance of its discovery.

It now seems that the discovery of the J/ψ was not only the crucial breakthrough in finding evidence for charm and helping to establish the marriage of the weak and electromagnetic interactions but also gave impetus to some very radical ideas about the nature of the strong interaction.

Charm: A testbed for quantum chromodynamics

The discovery by 't Hooft in 1971 that a self-consistent description of weak interactions could be made within a gauge theory framework stimulated a resurgence of interest in them. With the discovery of the predicted neutral currents in 1973 there was considerable excitement that pursuing the gauge theory idea might be the philosophers stone long-sought in attempting to crack the secret of the forces of nature.

As we have seen already in Chapter 7, the discovery that quarks carry three colours had enabled construction of an SU(3) gauge theory of their interactions—quantum chromodynamics (QCD). With the discoveries taking place in the weak interaction arena, interest in QCD as a theory of the remaining nuclear force rapidly grew.

In 1972 three independent groups of people, Gerhard 't Hooft, David Politzer and also David Gross and Frank Wilczek discovered an astonishing property of QCD. It implied that the attractive forces clustering quarks into protons and nuclei are not always strong. At *low* energies ($\simeq 1$ GeV) they are powerful but in *high* energy particle collisions ($\simeq 10$–100 GeV) the forces between quarks should be much more feeble.

Before the discovery of the J/ψ, the known hadrons all had masses in the 1 GeV region and the quarks experienced strong forces. However the charmed quark was some 3–5 times heavier than anything previously known and its discovery revealed an unknown high-energy world. The forces acting on the massive (energetic) charmed quarks are predicted by QCD to be weaker than the so-called strong forces familiar for the up and down quarks that form protons, neutrons, and nuclei.

A strong attraction between quark and antiquark causes them to mutually annihilate. Thus the ϕ meson built from a strange quark and its antiquark soon decays due to their mutual annihilation. The ψ meson is an analogous system but built from a massive charmed quark and antiquark rather than the lighter strange quarks. QCD theory predicts that the massive charmed pair are less strongly attracted to one another than are the strange pair. Thus they are less likely to mutually annihilate and so the ψ will survive much longer than the ϕ. This prediction by Politzer and Tom Appelquist was dramatically verified—the ψ lived almost 100 times longer than the ϕ. Thus the ψ discovery was pivotal in the development of our recent ideas; it confirmed the ideas of the electroweak theory that called for charm and by its elongated life showed that the QCD theory of the quark forces was also on the right track.

How many quarks?

Our everyday world can be thought of as being built from a hydrogen atom template (an electron and proton system) with the addition of neutrons (forming nuclear isotopes) and the existence of radioactivity where a neutron transmutes into a proton, emitting an electron and also a 'ghostly' neutrino. The electron and neutrino (leptons) appear to be structureless elementary particles. On the other hand, the proton and neutron are now believed to be clusters of up and down quarks and these quarks are apparently elementary particles along with the leptons.

This pair of up and down quarks has many properties in common with the lepton pair (made up from the electron and the neutrino), and there appears to be a deep connection between them (they all seem to have no internal structure, have spin $\frac{1}{2}$, respond to electromagnetic and weak interactions in similar ways, etc.). These lepton and quark pairs are today known as the 'first generation' of elementary particles.

For a reason that is not yet well understood, nature repeated itself. The muon (seemingly a heavy version of the electron) was discovered over a quarter of a century ago (p. 51). It too is partnered by a neutrino and shows no sign of any internal structure. Thus there is a 'second generation' of fundamental leptons (μ^-, ν_μ). We now know that there is a second generation of quarks also: the charmed and strange quarks are indeed siblings, connected to one another by the weak interaction so that the charmed quark transmutation into strange is analogous to the down to up transmutation that triggered neutron β-decay.

Most physicists had wondered why the muon existed, and whether there were more massive particles akin to the electron and muon. In particular Martin Perl at SLAC had repeatedly stressed the importance of searching for such entities and it was fitting that in 1975 his team found the first evidence for such a particle. Its mass is almost 2000 MeV (2 GeV), twice the mass of the proton and similar to the masses of the charmed mesons that were also discovered in the debris around that time. Experiments during 1976 in the Hamburg electron–positron machine confirmed this lepton's existence and helped to establish its properties. Just as the muon is a heavy version of the electron, so does this new particle, the 'tau' (τ), appear to be a yet heavier version of them both. It seems to be a structureless elementary particle, electrically charged and partnered by a neutrino (labelled ν_τ to distinguish it from the ν_e and ν_μ). This τ, ν_τ pair seems to be yet another repetition in nature, and acts like the electron and its neutrino, or again like the muon and its neutrino. Thus we have a 'third generation' of leptons.

As there is now a third generation of leptons, theorists argued that there should also exist a third generation of quarks to restore the elegant quark–lepton symmetry, Box 9.7. This new quark pair is labelled t and b

Box 9.7 Three generations of leptons and quarks

The charmed quark completes a second generation of particles. The discovery of the tau lepton and neutrino has established a third generation of leptons. The upsilon, found in 1977, revealed the first evidence for a bottom quark. A top quark is now being searched for to complete a third generation of quarks. Each flavour of quark occurs in any of three colours.

1st generation
Established 1960s.

2nd generation
Established 1974–76.

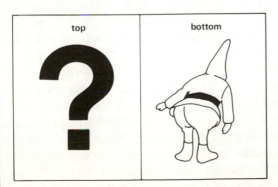

3rd generation (?)
Bottom found 1977.
Top predicted to exist: still being searched for. First tentative sighting claimed 1984.

for 'top' and 'bottom' (sometimes called 'truth' and 'beauty'). They were predicted to have electrical charges of $\frac{2}{3}$ and $-\frac{1}{3}$ just as was the case for the first generation (up, down) and the second generation (charm, strange).

If hadrons containing top or bottom quarks exist they must be much more massive than anything previously found, as no evidence for them had ever been found in low energy collisions of protons or of electrons and positrons. Just as the J/ψ meson is a bound state of $c\bar{c}$ (charmed quark and its antiquark) so analogous massive mesons Υ_b (upsilon) and Υ_t should exist made from $b\bar{b}$ and $t\bar{t}$ respectively. These Υ should be produced in the debris of proton–proton collisions (analogous to the way Ting and his collaborators discovered the J/ψ in 1974) and also in electron–positron annihilation (like Richter and his collaborators discovered the same object at Stanford that same year).

In the summer of 1977 a group of physicists led by Leon Lederman working at Fermilab near Chicago, discovered a massive Υ produced in proton–proton collisions. This Υ has a mass of 9.45 GeV, some three times as massive as the J/ψ and ten times the proton! Some properties of the production suggested that it was Υ_b ($b\bar{b}$) that had been found. This could best be tested by observing the way that the Υ was produced in electron–positron annihilations; however there was no machine available at that time with enough energy to be able to do this. (This is in marked contrast to the J/ψ whose discovery was announced simultaneously in proton–proton and electron–positron collisions.)

It was not until the summer of 1978 that the Hamburg electron–positron collider, DORIS, was able to produce the Υ. This was successfully achieved in August of that year, slightly over a year after its discovery in proton collisions at Fermilab.

The J/ψ had hidden charm, formed from a charmed quark and charmed antiquark binding together to form a meson with no net charm: particles with net charm were subsequently found. If the Υ has 'hidden bottom' (made from a bottom quark and its antiquark) then mesons built from a bottom quark with an up, down . . . antiquark should exist by analogy, Box 9.8. (These are somewhat misleadingly and irreverently said to possess 'naked bottom'.) Evidence for these emerged in 1980–81 at CESR (Cornell) and CERN, and some of their properties hint that the sixth quark, top, exists, though direct evidence of this is still awaited.

Thus there do indeed appear to be hints that nature contains a third generation of quarks partnering the third generation of leptons (τ, ν_τ). To really confirm this requires the discovery of Υ_t and mesons containing 'naked top' (or top mesons as they are sometimes called). The highest energies currently available in electron–positron annihilations have found no evidence for such a meson with mass less than about 30 GeV! Most theorists are confident that it exists, but no one yet has a well-accepted

Box 9.8 Bottom and top quarks: discovery and search

Just as J/ψ is made from charm and anticharm so is the upsilon, Υ, made from a bottom quark and a bottom antiquark.

The discovery of Υ in 1977 was the first example of a particle containing a bottom quark.

The 9.4 GeV mass arises from the two quarks each weighing 4.7 GeV.

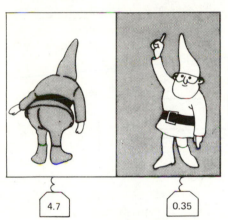

Mesons weighing just over 5 GeV containing bottom and up quarks are predicted to exist and some hints of their existence have emerged.

A sixth quark, top, is predicted. This will complete 6 quarks and half a dozen leptons.

The first manifestation of top quarks is expected to be the analogue of J/ψ and Υ: a meson containing top and antitop. All that we know so far is that if it exists it must be very heavy.

Box 9.9   Energy levels for excited states of molecules, atoms, a nucleus and a proton

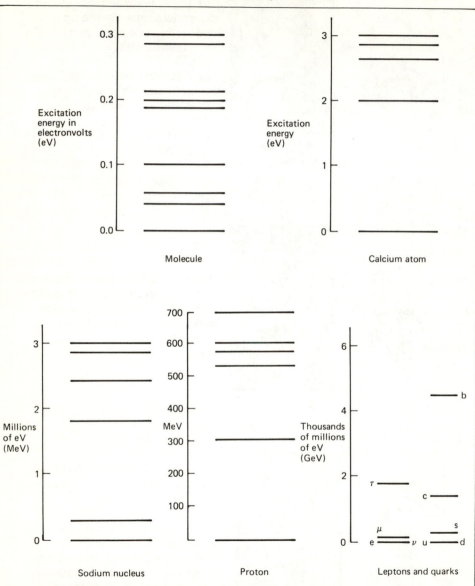

Note how the smaller is the structure, so the greater is the scale of excitation energy. The excitations correspond to rearrangements of the constituents. It is not yet known whether the quark and lepton spectrum is indicative of deeper structure in matter.

theory capable of predicting its mass. The attempt to discover particles containing a top quark or antiquark is one of the most important current projects experimentally. Their future discovery (or non-discovery?) will have dramatic significance for the development of our embryonic theories attempting to unify the weak, electromagnetic, and strong interactions, and also for intensifying the suspicion that quarks and leptons are intimately related to each other.

In a very real sense the recognition of the generation patterns for quarks and leptons is like the discovery of the Eightfold Way (Chapter 5) for hadrons and Mendeleev's periodic table for atoms (Chapter 2). The discovery of predicted top quarks would be analogous to that of the Ω^- discovery in the Eightfold Way and to the discovery of the elements gallium and germanium in the periodic table. Having confirmed the validity of the pattern the question that will then come to mind will be, 'what causes the pattern to exist?' In both the periodic table of the elements and the Eightfold Way for hadrons the cause was a deeper layer in the Cosmic Onion. In the case of quarks and leptons, however, no hint has emerged of an analogous substructure. It is suspected that their relationship and the cause of the generation patterns is more profound. Ideas are being developed but this problem is wide open at the moment. Refer to Box 9.9.

Postscript April 1985

⑩ *Grand Unification*

In the last three chapters we have seen how astonishing progress was made during the 1970s in understanding the fundamental building blocks and the forces that cluster them into observable matter. The current excitement goes beyond this, stimulated by patterns shared by the particles and the forces, which seem to be too clear cut to be mere chance. We may be getting here the first glimpse of a profound unity in nature that may have existed during the Big Bang epoch but which has become hidden during the expansion and cooling of the universe. It is only with the creation of local 'hot' conditions in particle collisions using high-energy accelerators that the glimpse of this one-time unity has been obtained. This chapter will describe the background to these ideas and also collect some other discoveries that have not entered the story so far. The final chapter will gather all these together to show how these insights in high-energy particle physics are giving a new understanding of cosmology and the history of the universe.

Coincidences?

Particles
The fundamental particles of all matter seem to be quarks and leptons. These all have spin $\frac{1}{2}$, obey the Pauli principle which prevents more than one particle of a given kind from being in any state, show no evidence of internal structure, and show a left-handed preference in weak inter-actions. In Chapters 8 and 9 we saw that the leptons and quarks form doublets which controls their response to the weak force. The weak force acts on the lepton doublets in *precisely* the same way as on quark doublets. As far as the weak interactions are concerned, leptons and quarks appear to be identical. Are quarks and leptons therefore not two independent sets of particles but instead related in some way? Is the fact that each lepton pair is accompanied by a quark pair the first hint, *á la* Mendeleev, of yet deeper structure?

146

Box 10.1   The puzzle of the electron and proton electrical charge: episode 2

On p. 108 we commented upon the profound puzzle of why the electron and proton have the same magnitude of charge. The charge of an electron was there suggested to arise by adding a discrete unit to a fundamental neutrino and that of a proton analogously from a fundamental neutron. While the former is possible the latter is not, now that we know that the neutron is composed of quarks.

The exciting discovery made recently is that the idea can be applied to leptons and quarks starting from the fundamental neutrino *in both cases* so long as we include colour en route!

Starting with the neutrino and removing one unit of charge gives the electron as before. If we added a unit of charge to the neutrino we would not obtain a known particle of matter (the e^+ is the positron). Suppose however that we admit the existence of three colours and paint this particle any of three ways, sharing the charge accordingly. Then we have red, yellow, or blue spin $\frac{1}{2}$ objects, each with charge $\frac{1}{3}$. This is nearly a quark; the up quark charge of $\frac{2}{3}$ and the down of $-\frac{1}{3}$ add to this very value.

So far, electric charge and colour (strong force) have entered the scene. The weak interaction connects ν^0 to electron, and down to up quark so the *difference* of up and down charges must equal that of the leptons. Starting from a charge of $\frac{1}{3}$ that is to be distributed between the up and down quarks only allows the solution: up $= +\frac{2}{3}$, down $= -\frac{1}{3}$. The electron and proton (uud) charges are then precisely balanced.

To obtain this result we have had to suppose two amazing things:

(i) Quarks and leptons are intimately related: in a sense quarks are leptons that have been given colour.
(ii) The strong (colour), and weak forces have to conspire with electromagnetic (charge) suggesting that they are not independent of one another.

The second question is still open but a possible answer to the first has emerged, surprisingly, as a result of insights into the nature of the forces (Chapters 7 and 8). To set the scene for this we should first not forget the electromagnetic and strong forces since here the leptons and quarks seem to behave quite differently—perhaps the weak interaction similarity was a red herring.

The quarks have charges $\frac{2}{3}$ or $-\frac{1}{3}$, the leptons have -1 or 0, and so the strength of coupling to the electromagnetic force differs for each. However it differs *only* in the overall scale of $\frac{2}{3} : -\frac{1}{3} : -1 : (0)$. Apart from this size effect all electromagnetic properties are identical, notably the ratio of magnetic moment and electrical charge is the same for electron and for up and down quarks (apart from inessential differences due to their different masses).

The place where a manifest difference appears is in strong interactions which quarks respond to whereas leptons are blind to them. The quantum chromodynamic theory relates this to the fact that quarks carry colour whereas leptons do not. However even this difference gives hints of a deeper unity: there appears to be a correlation between the existence of three colours and third fractional charges for quarks and the nonfractional charges of uncoloured leptons (see Box 10.1).

Forces
As recently as ten years ago the strong, weak, and electromagnetic forces seemed to be totally unrelated to one another, the strong force being a hundred times stronger than the electromagnetic, with the weak force yet another thousand times more feeble, and different varieties of matter being affected by these forces in quite different ways. But the perspective changed with the development of quantum chromodynamics and electroweak theories, and with the realisation that the apparent differences in strengths are manifestations of the forces only at low energies or at nuclear distances: one small region of nature's spectrum. Refer to Box 10.2.

Electromagnetism involves *one* (electrical) charge and quantum electrodynamics is a U(1) theory. The doublets of weak isospin that quarks and leptons form is a *two*-ness that is the source of an SU(2) theory similar to quantum electrodynamics. Combining these yields the SU(2) × U(1) theory of weak and electromagnetic phenomena. The *three* colours possessed by quarks, but not leptons, generate an SU(3) theory of quark forces, quantum chromodynamics, similar again to quantum electrodynamics. Thus there is a common principle at work, the essential difference being the one, two or three-ness—the U(1), SU(2) or SU(3) theory. This seems too much of a coincidence, and suggests that the three theories may be related.

A further hint of unification comes from the relative strengths of the three forces. The coupling strengths of the photon, W, and Z bosons and of the gluons, the force carries in the U(1), SU(2) and SU(3) theories, are numbers g_1, g_2 and g_3 respectively. (Conventionally $g^2/4\pi$ is denoted by α which $= 1/137$ for electromagnetic interactions, $\simeq 1$ to $\frac{1}{5}$ for coloured quark interactions in the proton at present accelerator energies. Although electromagnetic interactions are weaker than 'strong' interquark forces at present energies, we should remember the remarkable property of the SU(3) theory, noted on p. 139 that the interaction strength g_3 is not a constant but decreases as energy increases. Thus although strong at low energies (large distances), the interquark force weakens significantly at

Box 10.2   Do exotic forces operate at very short distances?

The nuclear force is strong but of such short range that its effects are feeble at atomic dimensions. If we were unable to probe distances of 10^{-12} cm we would be unaware of this force.

The 'weak' force of radioactivity is believed to be as strong as the electromagnetic force but is of extremely short range. To see its full strength we must probe distances of the order of 10^{-17} cm which requires head-on collisions of particles at energies of 100 GeV, something that is only just now becoming possible. Under these conditions, weak effects are as probable as electromagnetic. Studied from afar they appear to be more feeble.

This perspective raises the question of whether there are powerful forces at work on very short distance scales, far beyond the reach of present technology and whose effects are thereby hidden from us. The 'Grand Unified Theories' suggest that this is the case, in particular it is possible that forces act over a distance of 10^{-29} cm that will change quarks into leptons. To probe such distances requires energies of 10^{15} GeV, far beyond anything possible now or in the forseeable future.

However quantum mechanics comes to our aid here. When we say that the proton is 1 fm in size, we are making a statement about probability. About once in every 10^{30} years, the quarks in the proton will find themselves within 10^{-29} cm of one another and then this force can act on them, transmuting a quark into a lepton and causing the proton to decay, as described in the text. Thus we can study very short distance phenomena if we are patient and wait for the once in a blue moon configuration to happen. With enough protons we can tip the odds in our favour and need not wait too long. If we have 10^{30} protons available then we need only watch them for one year to have a good chance of seeing one decay.

higher energies. This property is not peculiar to SU(3) but occurs for all such theories, the only difference between SU(2) and SU(3) is the rate at which the weakening occurs as energy increases. For U(1) the force is predicted to get stronger at higher energies. Hence electromagnetism gets stronger at high energies and the strong force weakens. Extrapolating this to the huge energy of 10^{15} GeV all three forces turn out to be equally strong: α_{QED} has risen from 1/137 to about 1/42 and α_{QCD} has fallen from unity to this same 1/42 (Box 10.3). Thus (one divided by) 42 may indeed be the answer to the ultimate question.†

The forces are described by a common mathematics and have identical strengths at 10^{15} GeV. Under these conditions the idea of unification has

† See: Douglas Adams, *The Hitch Hikers Guide to the Galaxy* (Pan, 1980).

Box 10.3 Weak, electromagnetic, and strong forces change strengths at high energy

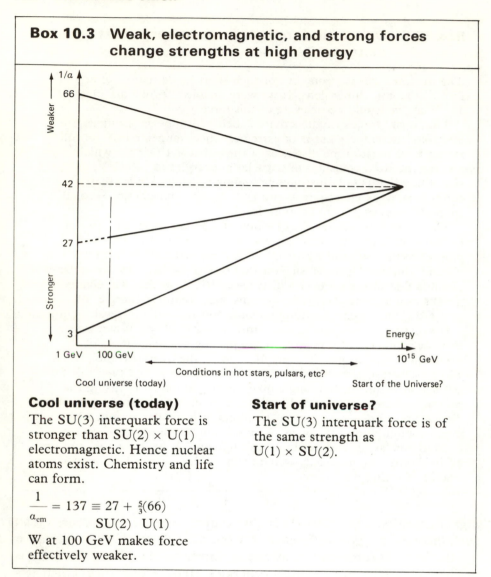

Cool universe (today)

The SU(3) interquark force is stronger than SU(2) × U(1) electromagnetic. Hence nuclear atoms exist. Chemistry and life can form.

$$\frac{1}{\alpha_{cm}} = 137 \equiv 27 + \tfrac{5}{3}(66)$$
$$\quad\quad\quad SU(2) \quad U(1)$$

W at 100 GeV makes force effectively weaker.

Start of universe?

The SU(3) interquark force is of the same strength as U(1) × SU(2).

some meaning; the unity is obscured at the cold low energies to which we had been restricted until the last decade. It is only very recently that the advent of accelerators capable of over 500 GeV energies has revealed evidence for the changing strength of the interquark forces. We are never likely to be able to create concentrations of 10^{15} GeV energies in the laboratory and see the full glory of unity of the forces, but if the hot Big Bang model of creation is right then head-on collisions at energies of 10^{15} GeV would have been abundant in that epoch. The current belief, which has exciting consequences for cosmology, is that in the high temperatures of the hot big bang there was indeed a unity among the natural forces with consequent production and decay mechanisms for particles that have been long forgotten as the universe cooled. Some of these processes appear to have been crucial in generating the universe that we see today. They will be described in the final chapter.

The idea that symmetrical behaviour occurs in phenomena at extreme temperatures but is obscured at lower temperatures is familiar in many contexts. One such is the transition from liquid to crystalline solid. At high temperature liquids are isotropic, all directions being equivalent to one another. As the temperature drops the liquid solidifies and may form crystals. These are not isotropic—the original full symmetry has been lost even though well-defined symmetries, less than total isotropy, remain. Magnets yield a second example—at high temperatures the atomic spins are isotropically randomly ordered and no magnetism ensues. At lower temperatures north and south magnetic poles occur.

Thus the notion that the universe possessed an innate symmetry at extremely high temperatures, such as at the Big Bang, which was obscured as it cooled to the present day is quite in accord with other physical experience.

How is the grand unification idea implemented, what are its consequences and how can we test it? These are the questions to which we now turn.

SU(5) and proton decay

In the $U(1) \times SU(2)$ gauge theory of electromagnetic and weak interactions, particles occur in pairs such as (e^-, ν_e), or (d, u). Emission or absorption of the γ and (W^+, W^-, Z^0) caused transitions from one member of a doublet to the other, thus:

$$e^- \leftrightarrow \nu_e + W^-$$
$$d \leftrightarrow u + W^-$$

In the $SU(3)$ theory of colour interactions the quarks occur in triplets such as (u_R, u_B, u_Y), or (d_R, d_B, d_Y). Emission or absorption of the coloured gluons cause analogous transitions within the triplet:

$$u_R \leftrightarrow u_B + g_{(RB)}$$

(the R − B denoting the 'red minus blue' gluon in this case), or

$$u_R \leftrightarrow u_Y + g_{RY)}$$

and so forth. There are eight $(= 3^2 - 1)$ gluons in total.

Box 10.4   The puzzle of the electron and proton electrical charge: episode 3

On p. 115 we saw how the electric charges of leptons or quarks were given by

$$\text{charge} = \pm \tfrac{1}{2} + \frac{Y}{2}$$

where Y was hypercharge and the $\pm \tfrac{1}{2}$ referred to the upper or lower member of the SU(2) multiplet. The doublets belong to an SU(2) × U(1) group structure; If Y = 0 then it would have been simply SU(2).

The sum of the charges of all particles in the multiplet is

$$\text{Total charges} = \frac{0}{\text{SU(2) piece}} + \frac{Y}{\text{U(1) piece}}$$

The vanishing of the charge sum in SU(2) is a feature that is common to any SU(N) group that contains photons as one of the bosons, and which conserves electrical charge.

If the strong, weak, and electromagnetic interactions are to be combined into a single grand SU(N) interaction then the sum of the charges of the leptons and/or quarks in the basic N-member family must vanish. This can be done in SU(8) where the 8 are the members of a generation of leptons, (e^-, ν_e) and of quarks (u, d) in each of red, yellow, or blue. In this case, the sum of lepton charges is -1 and that of quarks is $+1$, exactly balancing as required. Note how the charge and threefold colour conspire for quarks to balance the leptons charges.

It is possible to build an SU(N) theory without having N as large as eight. Most simply it is sufficient for N to be 5 and the five members are $(e^+, \bar{\nu}_e, d_R, d_Y, d_B)$ whose charges add to zero.

Notice that in either case it is necessary to have both quarks and leptons in the same family to yield a net zero of charge. Thus quark–lepton transmutations are to be expected. The SU(N) demand that the charges add to zero, is the mathematical formulation of the picture in Box 10.1, leading to the equality of positron and proton charges.

In a general SU(N) theory of particles and forces there will be N particles that can transmute one into the other and $(N^2 - 1)$ force carriers, refer to Box 8.5.

We can combine the U(1), SU(2) and SU(3) theories into a single 'unified' grand SU(5) theory. Here the *five* particles are, for example, the *two* leptons e^- and ν_e and the *three* colours of the antidown quark (each antidown quark has electrical charge of $\tfrac{1}{3}$ so the sum total of three antidowns and electron and neutrino is zero—required if charge is to be absolutely conserved in the theory—as observed in nature).

Box 10.5   SU(5)

The SU(5) theory contains a five and a ten particle family. The five took care of $\bar{d}_{R,B,Y}$, $[e^-, \nu_e]$. The ten contains $[d_{R,B,Y}, u_{R,B,Y}]$, $\bar{u}_{R,B,Y}$ and e^+ ($\bar{d}$ and $\bar{u}$ denoting the down and up antiquarks). The square brackets contain the pairs connected by the SU(2) weak interaction. Technically all the above refers to the left handed spin states of the fundamental particles.

The right handed SU(5) has as the five particle family the $d_{R,B,Y}$, $[e^+, \bar{\nu}_e]$ and the ten the $[\bar{d}_{R,B,Y}, \bar{u}_{R,B,Y}]$, $u_{R,B,Y}$ and e^-. There is no right-handed neutrino. Note that the theory contains precisely what appears to be needed and no more. If neutrinos have masses and both left- and right-handed varieties exist, then this scenario will be incomplete.

The leptons and quarks of the first generation (p. 141) are forced on us (why nature chose to repeat this once, and probably twice more, is not understood).

The threefold colour and the fractional charge of the down quark go together to balance the electron charge. The weak interaction, subsumed in the theory, requires the up quark with charge $\frac{2}{3}$ and neutrino with zero charge in order that the $W^\pm$ bosons can interact with the down quark and electron. Thus the astonishing equality of proton and positron (anti-electron) electrical charges is forced upon us.

The 24 force carriers consist of the photon, W^+, W^-, and Z^0 of the electromagnetic and weak interactions, the eight gluons of the colour or strong interaction and twelve new particles. These are called X and Y particles and the theory requires them to carry any of three colours and have electrical charges of $+\frac{4}{3}$ or $-\frac{4}{3}$ and $\frac{1}{3}$ or $-\frac{1}{3}$. They will transmit a new force which can cause quarks to change into leptons, e.g.

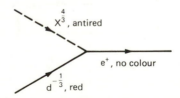

or quarks into antiquarks as in

$$\begin{cases} \qquad\quad u \to X + \bar{u} \\ \text{charge} \quad \tfrac{2}{3} \to \tfrac{4}{3} + (-\tfrac{2}{3}) \end{cases}$$

Box 10.6   The range of forces

The uncertainty principle (p. 16) showed the typical range that a particle of mass m can transmit a force to be $r = h/mc$. The magnitude of this as a function of mass is shown in the table.

mass m	distance s	effect
pion $\simeq \frac{1}{10}$ GeV	10^{-12} cm	strong at nuclear size $\simeq 10^{-12}$ cm
W boson $\simeq 10^2$ GeV	10^{-15} cm	weak at nuclear size
X boson $\simeq 10^{16}$ [GeV] (?)	10^{-29} cm	feeble for proton size: generates proton decay (?)

Massless particles can transmit forces over infinite distances. The massless graviton leads to an infinite range gravitational force. The massless photon leads to an infinite range electromagnetic force. However, the existence of positive and negative charges which cluster to neutral in practice tends to cut off the effective range of the force. Gluons are massless and similarly might be expected to generate an infinite range force. However, the colour clustering to neutral appears to be absolute and its range is cut off analogously to the electromagnetic case. The gluons themselves have colour and so cluster to neutral glueballs. The lightest of these is believed to be about 1 GeV in mass, and so the colour forces are not transmitted beyond the proton size. Quarks are coloured and their colourless clusters have the pion as the lightest example. Thus the pion can be transmitted the furthest of all hadrons and is the important contributor to the nuclear force.

The X particles can generate a force between the quarks inside a proton that will cause the proton to decay. If we put the two above processes together inside a proton we start with uud (proton) and end up with $u\bar{u}e^+$ (a π^0 and positron).

If the X boson weighs around 10^{15} GeV, then it can only transmit forces over a distance of order 10^{-29} cm (Box 10.6). The proton's size is 10^{-12} cm, some seventeen orders of magnitude larger than this, and so it is exceedingly rarely that the quark–lepton transmutation can occur in the proton. Thus, in practice, the probability that a proton will decay is correspondingly small—calculations suggest that its lifetime is some 10^{30} years.

As the lifetime of the universe is 'only' 10^{12} years then this prediction might appear to be only of academic interest. However it is possible to test it. If one has enough protons available (such as in a large tank of water) then you may be fortunate enough to see one decay in a year, say. Large 'swimming-pools' have been built deep underground to eliminate spurious effects due to cosmic rays (they are continually bombarding the Earth's surface but are absorbed by solid material and so do not penetrate far underground). Deep mines are one possibility and in Europe the Mont Blanc tunnel (on average 3000 metres below 'ground') has been suggested. The pool is surrounded by detectors—and then you wait patiently!

There are many technical problems to be overcome in designing and preparing such experiments. However the stakes are high, and in two or three years the first results might begin to emerge. The first experiments are finding some tantalising phenomena which might indeed be due to protons decaying, but their interpretation is still controversial. News on this is something to look out for.

This is but one example of new processes predicted by grand unified theories to be important at high energies, which may have played a major role in the development of the early hot universe. Under such hot conditions quarks could frequently have interacted at less than 10^{-29} cm range and so the quark–lepton transmutation could have been a common feature of that epoch and have played an important role in establishing the dominance of matter over antimatter in the universe. We cannot reproduce these hot conditions in the laboratory. However, at low energies, to which we are limited, these processes still occur but exceedingly infrequently. Thus the grand unified theory can be tested even at low energies if one waits long enough or studies enough material.

Here is a more extreme example of the case of the weak interaction transmitted by W bosons of order 100 GeV mass. In head-on collisions of particles above 100 GeV energies, the electromagnetic and weak interaction process will be equally likely to occur. However we don't have to go all the way to 100 GeV collisions to study weak interaction effects. These still occur at low energies albeit less frequently, hence its naming as 'weak' interaction, as a result of our low energy historical experiences. So it is with the X bosons of 10^{16} GeV: at low energies their effects are feeble but not totally absent.

Theorists are now trying to think of low energy manifestations of X particles which can be looked for in experiments with a view to confirming the grand unified theory. This is likely to be an active line of research in the 1980s.

Although the complete unification is predicted to happen only at 10^{15}–10^{19} GeV, the coming together of the electromagnetic and weak interactions is expected at 10^2 GeV and has recently become amenable to study at CERN. In the past, accelerators have been restricted to energies where the weak force is feeble in its effects, in particular the weak radiation—W and Z particles—cannot be produced whereas electromagnetic radiation—photons—is produced copiously. The W and Z bosons can only be produced at energies where the electromagnetic and weak forces have become as one. There is intense activity at CERN, attempting to produce W and Z bosons, studying their behaviour and, hopefully, directly verifying the uniting of two of nature's forces once the cold, low-energy world is left behind. Two hundred W bosons and a score of Z bosons had been produced by January 1985. Everything so far suggests that the weak and electromagnetic interactions are united at 10^2 GeV. The next step will be to produce the Z and W in abundance to see if this unity is indeed as predicted or whether new surprises are in store. If all goes well then we will have obtained the first proof that the cold universe of today exhibits multifarious phenomena that are direct consequences of the coldness: in hot conditions unification occurs.

Supposing that all these things come to pass and that the grand unified theory is at least a shadow of reality, then processes rare at room temperature should occur abundantly at 10^{15} GeV. Particle physicists and cosmologists are now re-examining their theories of the post big bang epoch. The new processes predicted by grand unified theories could have been important then and have played an important role in creating our present universe. Are there relics of such reactions present today? Can grand unified theories provide clues to some of the cosmological puzzles still plaguing the big bang theory?

These questions are currently being actively investigated. There are some exciting indications that reactions described in grand unified theory were indeed important in the early epochs. This brings cosmology and particle physics together and is the theme of our final chapter.

11 Cosmology, particle physics, and the Big Bang

The modern view of a large-scale universe consisting of island galaxies each with billions of stars, stems from the work of the American astronomer Edwin Hubble. In 1924 he showed that galaxies exist beyond our own Milky Way, a discovery that indicated the universe to be far far greater than had been previously imagined. The more astonishing discovery, announced in 1929 and confirmed in subsequent decades of astronomical research, was that these galaxies are all rushing away from one another. The further away from us the galaxies are, the faster they are receding, which shows that the universe is expanding in all directions.

This discovery gave a clue to the possible origin of the universe, since if the universe is expanding now, then it must have been much smaller in the past. From the observed rate of expansion we can extrapolate back to that past and we find that some 10 to 20 thousand million years ago the galaxies were gathered together in a single mass, their material compressed into an infinitesimal volume. This singular state of affairs is what we currently call 'the start of the universe'.

In 1965 Arno Penzias and Robert Wilson discovered black body radiation whose intensity is the same in all directions in space implying that it has to do with the universe as a whole. The existence of a uniform thermal radiation existing throughout an expanding universe has remarkable consequences. As the universe expands so will the thermal radiation cool down, its absolute temperature halving as the radius doubles. Conversely, in the past when the universe was much smaller, so the thermal radiation was correspondingly hotter. Its present temperature is $3°$ above absolute zero; extrapolating back 10 thousand million years reveals that the early, extremely dense universe was incredibly hot.

These observations all agree with the 'Hot Big Bang' model of the birth of the universe suggested as early as 1948 by Alpher, Gamow, and Herman, and now so well established by data that it is often referred to as 'The Standard Model'.

Thus we live not in a static unchanging universe but one that is developing in time. The realisation that the infinite variety in our present

157

universe has evolved from that early cauldron has provided a totally new perspective on astrophysics and high energy physics. Some of the diversity in the laws governing the forces that act within and between atoms is now thought to arise as a result of this evolution. Grand Unified Theories (Chapter 10) imply that natural forces, and the particles on which they act, exhibit a uniformity at extreme energies which is obscured at low energies. Thus our long-standing low energy experience may be only a partial guide in constructing theories of the hot early universe. Grand Unified Theories paint a picture of a new-born universe in which exotic processes were commonplace but which have been long since hidden from view. This new theory has provided fresh insights into the workings of the early universe and has brought the previously disparate subjects of high energy physics and astrophysics together.

The universe today

All of our scientific research has been performed during a small spot of time in the vast life of the universe. Astronomers study distant galaxies whose light started its journey to Earth millions of years ago. From these observations we can, in effect, study the universe of long ago and it seems that, to the highest accuracy available, the laws of physics applicable today are the same as operated millions of years ago.

However, we cannot see back into the first million years of the universe by observations with optical or radio telescopes, or by other electromagnetic means. The background radiation originated in that epoch and is a veil beyond which our eyes cannot see. However, there are relics of earlier epochs such as the abundances of hydrogen relative to helium in the universe, the densities of matter relative to radiation photons and, closest to home, the existence of us—matter to the exclusion of antimatter. These are the observable signals, the beacons that show what occurred in that hidden first million years. Clues as to why they are as they are have emerged from high-energy particle collisions where, in the laboratory, local hot conditions are created that imitate the state of that early universe.

Most of the matter in the universe as viewed today is in stars which shine by burning light elements and making heavier ones. These processes have been well understood for nearly a quarter of a century and calculations imply that all of the nuclei of the atoms heavier than lithium, in particular the carbon in our bodies, the air in our lungs and the earth beneath our feet, were cooked in the nuclear furnaces of stars relatively recently.

These stars are collected into galaxies which in turn are gathered into clusters. Hubble's observations showed that these galactic clusters are moving apart from one another at a rate of about 30 kilometres per second for every million light years that they are currently separated. The known matter density is some two to ten times too low to provide enough gravitational pull to slow the universe's expansion to a point where it will eventually collapse back inwards. An open question at present is whether the new infra-red astronomy techniques will reveal concentrations of previously undiscovered cold stars that will bring the matter density up to the critical level. Another possibility, suggested by grand unified theories, is that neutrinos are massive, in which event they could provide enough 'missing matter' to close the universe.

Even if there is a factor of 2, 5, or even 10 underestimated in the density of matter, there is no doubt that the number of baryons is swamped by the number of photons in the background radiation. The number density of neutrons and protons in the universe as a whole is about 10^{-6} per cm^3, whereas the density of photons in the background radiation is of the order of 500 per cm^3 leading to a very small ratio of baryons to photons

$$\frac{N_B}{N_\gamma} \simeq 10^{-9 \pm 1}$$

This number has long been a puzzle. One's immediate guesses would suggest a ratio either near to unity or much nearer zero. If the universe contains matter and no antimatter then why are photons 10^9 times more abundant than baryons? The rational explanation would be that they were produced from annihilation of matter and antimatter. In the primordial fireball of the big bang, the energy density was so great that matter and antimatter were continuously created and destroyed. Why did the destruction leave such a 'small' excess of photons—why did it leave an excess of matter? These questions have been unanswered and unanswerable until the advent of the grand unified theory.

The history of the universe with some important dates is shown in Box 11.1. Its development from the first seconds is described in Weinberg's *The First Three Minutes* so I shall only briefly comment on this, and emphasise the new insights from grand unified theory and the push-back of our understanding to within 10^{-33} seconds of genesis.

Five stages in a cooling universe

In the history of the universe, summarised in Box 11.1 we can identify five epochs of temperature. The first is the ultra-hot of 10^{32} K during the first 10^{-43} seconds where quantum gravity effects are strong. We have little idea how to describe this epoch mathematically, a satisfactory quantum theory of gravity is still awaited and until it is formulated we shall remain ignorant on the dynamics.

Box 11.1 Important dates in the history of the universe

Time	Temperature	Typical energy	Possible phenomena
10^{-43} s	10^{32} K	10^{19} GeV	Gravity is strong. Quantum gravity theory required (not yet available).
10^{-37} s	$>10^{29}$ K	$>10^{16}$ GeV	Strong, electromagnetic, and weak forces united.
10^{-33} s	10^{27} K	10^{14} GeV	Processes controlled by superheavy bosons start to freeze. Predominance of matter over antimatter probably established.
10^{-9} s	10^{15} K	10^2 GeV	Production of W bosons starts to freeze. Weak interaction weakens relative to electromagnetism. (Maximum energies attainable by latest high energy accelerators.)
10^{-2} s	10^{13} K	1 GeV	Colour forces acting on quarks and gluons, cluster them into 'white' hadrons. Colour forces hidden from view. Protons and neutrons appear.
100 s	10^9 K	10^{-4} GeV $= \frac{1}{10}$ MeV	Nucleosynthesis—helium and deuterium created.
10^6 years	10^3 K	1/10 eV	Photons decouple from matter. Background radiation originates (optical and electromagnetic astronomy cannot see back beyond this epoch).
10^{10} years	3 K Background radiation. Stars provide local hotspots.	10^{-3} eV	Today. Galaxies and life exist.
$\lesssim 10^{32}$ years			All matter erodes away if no gravitational collapse of the universe. If the latter big crunch, then may be followed by new big bang. If so, does the whole cycle repeat?

Einstein showed in his general theory of relativity that space-time and gravitational forces are profoundly related. In consequence, it is not clear what 'time' means when gravitational forces are singularly strong. Thus the notion of the 'start of the universe' is ill-defined. We can extrapolate back from the present time to the point where the above unknowns appear, then if we naively project back still further we find that time-zero was 10^{-43} seconds earlier. However it is not clear what, if anything, this has to do with the 'start of the universe'. With this caveat applied to the first entry in Box 11.1 we can redefine all subsequent times to date from 'time that has elapsed since gravitational forces weakened'.

The hot epoch is a period of isotropy and no structure. At temperatures above 10^{29} K the typical energies of over 10^{16} GeV are so great that the fireball can create massive particles, such as X and $\bar{X}$, as easily as light quarks, leptons, and their antiparticles. The strong, weak, and electromagnetic interactions have equal strength, quarks and leptons readily transmute back and forth and a unity abounds that will soon be lost forever.

The warm epoch covers 10^{29} to 10^{15} K, where typical energies are 10^{16} to 10^2 GeV. This period of cooling is the one where first the strong and then weak and electromagnetic interactions freeze out and go their separate ways. It is particularly exciting to us that the lower end of this energy range has recently become accessible to laboratory 'high-energy physics' experiments at CERN. The symmetry between the weak and electromagnetic interactions is only broken at the very cold end of this epoch, so its restoration in the laboratory is eagerly awaited. Needless to say, if it is seen to be so restored then this will give theorists confidence that this whole scenario is more than science fiction. The discoveries of the W and Z are major steps in this regard.

Next, we come to the cool epoch. It is now so cool relative to 10^2 GeV that the 'weak' force has indeed become weak. Electric charge and colour are the only symmetries remaining in particle interactions.

The physics of this cool epoch has been studied in high-energy physics experiments during recent decades. At the cold end of this region the coloured objects have clustered into white hadrons and are hidden from view. We are blind to the colour symmetry and the strong interactions take on the form that binds the atomic nucleus and has been known for 50 years. This brings us to the cold epoch—the present-day conditions on Earth.

We now have the very antithesis of the isotropy present in the hot epoch. The different strengths in the natural forces and the various families of particles, nuclei, electrons, and neutrinos, upon which the forces act in differing ways, give rise to rich structure in the universe. Galaxies, stars, crystals, you and I, all exist.

Light elements and neutrinos

The heavy elements that constitute the bulk of the material on Earth had their nuclei formed inside stars. There, light nuclei fuse together producing heavy nuclei and releasing the energy by which the stars are visible. If we can explain the existence of heavy elements in terms of light ones, then how were the light ones—hydrogen and helium—created?

The hydrogen and helium in the universe today were produced about three minutes after the Big Bang and are a relic of that time. The temperature then was about 10^9 K and 'nucleosynthesis' (formation of nuclei) occurred very rapidly from the abundant neutrons and protons because these can fuse to form deuterium without being immediately ripped apart by hot photons: 'photodisintegration'. The temperature balance is very critical—cool enough that there is no photodisintegration, yet hot enough that two deuterons can collide violently and overcome their electromagnetic repulsion (they each carry positive charge due to their proton content).

All the neutrons form deuterium and all the deuterium forms helium–4:

$$n + p \rightarrow d + \gamma; \quad \gamma + d \nrightarrow n + p$$
$$d + d \rightarrow He^4$$

Small amounts of helium–3 and lithium–7 were synthesised at the same time but production of heavier elements was prevented because there are no stable isotopes with atomic mass 5 or 8.

All the neutrons and a similar number of protons have formed light nuclei; excess protons remain as hydrogen. Thus the ratio of helium to hydrogen present today tells us the neutron–proton ratio at nucleosynthesis. Astrophysicists can compute this ratio and it turns out to be rather sensitive to the precise rate at which the universe was expanding. This is controlled by the density of matter and also by the number of light particles such as neutrinos. The observed He^4 abundance today includes a small amount that has been produced in stars subsequent to the first three minutes. Allowing for this it seems that there can be at most 3 or 4 varieties of neutrino in nature ($\nu_e, \nu_\mu, \nu_\tau \ldots$?).

We have seen how particle physics has uncovered a correspondence between lepton pairs and quark pairs ('generations'—Chapter 9). At present we have evidence for possibly three generations: (up, down), (charm, strange), and (top?, bottom) quarks, (ν_e, e$^-$), (ν_μ, μ^-), and (ν_τ, τ^-) leptons. The helium abundance constraining the neutrino varieties to 4 would imply that at most one more generation exists. Thus do studies of the big bang constrain particle physics (and conversely).

Quarksynthesis and the predominance of matter

The origin of heavy elements has been explained in terms of light ones. The light ones were formed by fusion of neutrons and protons, and these were formed by the clustering of coloured quarks into threes with no net colour.

These quarks were produced in the original fireball where radiation energy was converted into quarks *and an equal number of antiquarks*. Thus the genesis of the ingredients of our present universe is understood, but how was this excess of quarks (and so matter) generated?

If one starts off with a net quark number of zero (i.e. the number of quarks minus antiquarks) then an excess of quarks over antiquarks can only occur if interactions exist that do not preserve net quark number. Grand unified theories do indeed contain such interactions. As outlined in Chapter 10, two quarks can convert into an antiquark and an antilepton which changes the net quark number by -3 and, in particular, causes protons to decay.

The X bosons that are believed to be responsible for generating this decay are extremely massive. Even so, very early in the big bang the temperature was so great that massive X particles and $\bar{X}$ antiparticles were copiously produced in equal amounts. As the universe expanded and cooled, the production of very massive material became disfavoured and soon ceased totally. Any X and $\bar{X}$—that had not collided and mutually annihilated then decayed into pairs of quarks or quarks and leptons:

$$X \rightarrow q + q \quad \text{or} \quad \bar{q} + \bar{l}$$
$$\bar{X} \rightarrow \bar{q} + \bar{q} \quad \text{or} \quad q + l$$

At first sight this does not seem to help. It is true that an X decay produces quarks and antiquarks asymmetrically, but the $\bar{X}$ decay appears to restore the balance. However, there is no reason why the fraction of X particles that decays into quark pairs has to equal the fraction of $\bar{X}$ that decays into antiquarks. Hence even though we have an equal number of X and $\bar{X}$ to begin with, we can end up with an excess of quarks over antiquarks. See Box 11.2.

Not only can this happen in principle but there are reasons to suspect that it does in practice. James Cronin and Val Fitch won the Nobel Prize for their part in the 1964 discovery that such matter–antimatter asymmetry does occur, at least in K meson decays. Starting with an equal mixture of K^0 and $\overline{K^0}$, their decays into $e^+ \pi^- \bar{\nu}$ and $e^- \pi^+ \nu$ are matter–antimatter correspondents.

$$e^+ \equiv \overline{e^-}$$
$$\pi^- \equiv \overline{\pi^+}$$

Yet the two sets of products are *not* equally produced: the decay $K^0 \rightarrow e^+ \pi^- \bar{\nu}$ is about 7 parts in a thousand more frequent than $\overline{K^0} \rightarrow e^- \pi^+ \nu$. An asymmetry between matter and antimatter that has been observed for equal mixtures of K^0 and $\overline{K^0}$ leads one to expect an analogous asymmetry for an equal mix of X and $\bar{X}$. The dominant decays

---

Box 11.2   Creation of matter

Early hot Big Bang
Extremely energetic photons produce equal
numbers of X and $\bar{X}$. $\Big\}$ $\gamma \to X + \bar{X}$

Later cooler conditions (10^{-33} s)
Suppose X decays into quark pairs fraction a
of the time and into antiquark and antilepton $\Big\}$ $X \to a(qq) + (1 - a)(\bar{q}\bar{l})$
the rest.

The $\bar{X}$ correspondingly decays into antiquark
pairs or quark–lepton fractions b and $(1 - b)$. $\Big\}$ $\bar{X} \to b(\bar{q}\bar{q}) + (1 - b)(ql)$

If a and b are identical then there is matter–
antimatter symmetry but if a > b then matter
dominates.

$$a = b : \begin{cases} N_q = N_{\bar{q}} \\ N_l = N_{\bar{l}} \end{cases}$$

$$a > b : \begin{cases} N_q > N_{\bar{q}} \\ N_l > N_{\bar{l}} \end{cases}$$

The minority q, l, are annihilated by $\bar{q}$ and $\bar{l}$
This produces radiation. $\Big\}$ $N_\gamma \sim N_{\bar{q}} + N_l$

The excess q and l form the material from
which our present universe was built. $\Big\}$ $N_B \sim N_q - N_{\bar{q}}$

Today
The 3 background radiation is the remnant
of the photons produced in the annihilation
and is a measure of the number of antiquarks
and leptons. From observation we see that
matter is only a small remnant of what was
originally produced. $\Bigg\}$ $\dfrac{N_B}{N_\gamma} \sim 10^{-9}$

---

produce quarks and leptons slightly more than antiquarks and antileptons, leading to a small excess of matter over antimatter.

Later, all the antiquarks annihilated with quarks and all the antileptons
annihilated with leptons. The excess quarks and leptons that remained
formed our universe: most of the antimatter did not survive the first
millionth of a second.

Exodus

Will the universe expand forever, or is there enough matter in it to provide sufficient gravitational pull that it will eventually collapse under its own weight?

The first possibility presents a depressing outlook. Grand unified theory predicts that all matter will decay with a half life of about 10^{30} years. Thus all matter will erode away and the universe will end up as expanding cold radiation. There is a growing suspicion that the second possibility, collapse, is a more likely eventuality and that neutrinos may be the agents responsible.

In the primordial fireball neutrinos should have been produced as copiously as were photons. Thus there should exist about 10^9 neutrinos per proton. Consequently if an individual neutrino weighed as little as 1–10 eV then the bulk of the mass in the universe would be carried by neutrinos, not baryons. As stars and other observed matter are built from the latter, there could be (at least) twice as much mass in the universe as we are currently aware of and this could be sufficient to lead to ultimate gravitational collapse.

Do neutrinos weigh an electronvolt or more? No one yet knows. There is no known principle that requires them to be massless and, in contrast, grand unified theories tend to predict that they will have small masses of this sort of magnitude. There are some anomalies in the rate at which neutrinos arrive from the sun, and some as yet unreproduced controversial experiments that suggest that neutrinos *do* carry mass. This is an active experimental programme at present.

If neutrinos are found to have mass then this will be important for grand unified theory and for astrophysics. There are some puzzles about galaxy formation that could perhaps be answered if neutrinos have mass. Furthermore the gravitational attraction of massive neutrinos could cluster them together forming halos around galaxies, stabilising the galaxies during their formation and leading to observable effects on galactic interactions and motions. But to return to our original observation, the most dramatic long-term effect would be that the extra gravitational attraction generated by these massive neutrinos could be sufficient to slow down the universal expansion to the point where it stops, the universe contracts and then expires in a big crunch.

And after the big crunch—a new big bang and so on *ad infinitum*?

Here we have reached the frontier of speculation based on our current knowledge, the question just posed is currently metaphysical. But metaphysics of yesteryear has become amenable to scientific investigation today as a result of our recent advances. Future discoveries may enable us to ask and answer questions that are still unimaginable.

It is less than a hundred years since Becquerel discovered the key that unlocked the secrets of the atom and led to our current all-embracing vision. Where will another century take us? For all the wonders that our generation has been privileged to see, I cannot help but agree with Avvaiyar:

> What we have learned
> Is like a handful of earth;
> What we have yet to learn
> Is like the whole world.

Suggestions for further reading

These do not necessarily represent the best literature available, but they are all readable and I have enjoyed them.

The story of uranium, development of atomic and nuclear physics

Bickel, L. *The Deadly Element*, Macmillan, 1980.
Fermi, L. *Atoms in the Family*, Chicago, 1954. Fermi's life story.
Bohr, Niels. *Collected Works vol. 9. Work on Nuclear Physics*, North Holland, 1983.
Snow, C. P. *The Physicists*, Macmillan, 1981.

More about particles and forces

Asimov, Isaac. *The Neutrino*, Dobson, 1966.
Davies, P. C. W. *The Forces of Nature*, Cambridge University Press, 1979.
Polkinghorne, J. C. *The Particle Play*, Freeman, 1979.
Kaufman, W. (editor). *Particles and Fields*, Scientific American reprints, Freeman, 1980.

Modern ideas on particles and astrophysics

Weinberg, S. *The First Three Minutes*, Deutsch, 1977.
Mulvey, J. (editor). *The Nature of Matter*. Oxford University Press, 1981. A collection of popular lectures including ones by Murray Gell-Mann and Abdus Salam.
Bath, G. (editor). *The State of the Universe*, Oxford University Press, 1980.

For readers wanting more on quarks, up to a professional level:

Close, F. E. *Introduction to Quarks and Partons*, Academic Press, 1979.

Useful quantities and their symbols

Some physical constants

Planck's constant

$$\hbar = h/2\pi = 6.6 \times 10^{-22}\,\text{MeV s}$$
$$= 1.05 \times 10^{-20}\,\text{J s}$$

Velocity of light
$$c = 3.0 \times 10^8\,\text{m s}^{-1}$$

Charge of proton
$$e = 1.6 \times 10^{-19}\,\text{C}$$

'Fine structure constant'
$$\alpha = e^2/\hbar c = 1/137$$

Energy and mass

Electron volt

$$1\,\text{eV} = 1.6 \times 10^{-12}\,\text{erg} = 1.6 \times 10^{-5}\,\text{J}$$
$$1\,\text{keV} = 10^3\,\text{eV};\ 1\,\text{MeV} = 10^6\,\text{eV};$$
$$1\,\text{GeV} = 10^9\,\text{eV}$$

Electron mass
$$m_e = 0.511\,\text{MeV}/c^2 = 9.1 \times 10^{-28}\,\text{g}$$

Pion mass
$$m_\pi = 135\,\text{MeV}/c^2\ (\pi^0);$$
$$140\,\text{MeV}/c^2\ (\pi^+ \text{ or } \pi^-)$$

Proton mass
$$m_p = 938.3\,\text{MeV}/c^2$$

Neutron mass
$$m_n = 939.6\,\text{MeV}/c^2$$

Measures of strength of natural forces

Newton constant of gravitation
$$G = 6.67 \times 10^{-11}\,\text{Nm}^2\,\text{kg}^{-2}$$

Fermi constant of weak force
$$G_F \approx 10^{-5}\,(\text{GeV})^{-2}$$

Fine structure constant
$$\alpha = e^2/\hbar c = 1/137$$

Ratio of gravitational to electromagnetic forces in hydrogen
$$\approx 10^{-40}$$

($10^{-40} \approx$ ratio of proton and universe's radii)

Some characteristic lengths and times for light travel

quantum gravity
$$\sqrt{Gh/c^3} \approx 10^{-35}\,\text{m} \to 10^{-43}\,\text{s}$$

nuclear force
$$\hbar/m_\pi c = 10^{-15}\,\text{m} \equiv 1\,\text{fm}$$
$$\equiv 1\,\text{fermi} \to 10^{-23}\,\text{s}$$

atomic dimensions
$$\hbar/m_e c \approx 10^{-10}\,\text{m} \to 10^{-18}\,\text{s}$$

maximum distance travelled by light since start of universe
$$10^{10}\,\text{years at } 3 \times 10^8\,\text{ms}^{-1} \to 10^{26}\,\text{m}$$

Glossary

a, a_s, a_1, a_2, a_3	See **coupling constant**.
a-**particle**	A helium nucleus.
Angular momentum	A property of rotary motion analogous to the familiar concept of momentum in linear motion.
Antimatter	For every variety of particle there exists an anti-particle with opposite properties such as sign of electrical charge. When a particle and its anti-particle meet they can mutually annihilate and produce energy. Thus **antiquark, anti-proton**, etc.
Atom	System of **electrons** orbiting a **nucleus**. Smallest piece of an element that can still be identified as that element.
β-**decay**	Decay of a **radioactive** nucleus with production of an electron (β-particle). The underlying process is the transmutation of a neutron into a proton with electron and neutrino produced as a consequence. This process is controlled by the weak interaction and was its first known manifestation.
β-**particle**	Electron emitted in radioactive decay of a nucleus (β-decay).
Baryon	Nuclear particle, e.g. proton, built from three quarks.
Big Bang	The galaxies are receding from one another: the universe is expanding. The Big Bang theory proposes that this expansion began 10 to 20 thousand million years ago when the universe was in a state of enormous density.
Black body radiation	A hot black body emits radiation with a characteristic distribution of wavelengths and energy densities. Any radiation having this characteristic distribution is called black body radiation. Any system in **thermal equilibrium** emits black body radiation.

Boson   Generic name for the carriers of forces, such as the photon, gluon, and W particle for the electromagnetic, colour, and weak forces.

Bottom   A flavour of quark. The *bottom quark* like **strange** and **down** quarks has electrical charge $-\frac{1}{3}$ and is the heaviest quark currently known.

Cathode ray tube   Prototype television type.

Charm   Property of matter possessed by all material containing a *charmed quark*. This quark has electrical charge $+\frac{2}{3}$.

Charged current   Weak interaction in which the electrical charges of the participating particles get swapped around—typically involves a neutrino transmuting into a charged lepton or vice versa. Contrast **neutral current**.

Colour   Property possessed by quarks and gluons. A 'threefold type' of charge akin to electrical charge, believed to be the source of the strong force between quarks and described by the quantum chromodynamic theory of the strong interaction. Colour is absolutely neutralised at the elementary particle level of the Cosmic Onion.

Conservation   If the value of some quantity is unchanged throughout a reaction the quantity is said to be conserved.

Cosmic rays   High energy particles and nuclei coming from outer space.

Cyclotron   Early type of particle accelerator.

Deuteron   Nucleus comprising one proton and one neutron, labelled D. Sometimes called heavy hydrogen, and occurs in heavy water D_2O, contrast H_2O.

Down quark   Lightest quark with charge $-\frac{1}{3}$. Constituent of proton and neutron.

Eightfold Way   Classification scheme for elementary particles established *circa* 1960. Forerunner of quark model.

Electromagnetic radiation See Box 2.2 on page 12.

Electromagnetism	One of the forces of nature. Associated with the exchange of photons by particles that contain electrical charge. Holds atoms together. More details in Box 4.1 on page 34.
Electroweak force	Modern name for force based on a theory that combines electromagnetic and weak forces.
Electron	Negatively charged elementary particle constituent of atoms. When exchanged between atoms, leads to chemical combination. Carrier of electricity through wires.
Electronvolt (eV)	Unit of energy. Typically 1–10 eV is the amount of energy per atom involved in chemical reactions. 1 eV is the energy gained when an electron is accelerated by a potential of one volt.
Elementary particle	Once thought to be basic building blocks of matter—see **leptons** and **hadrons**. Leptons (like electrons) are today still believed to be truly fundamental. Hadrons (strongly interacting nuclear particles), like the proton, are built from more fundamental entities called **quarks**.
Feynman diagram	Pictorial representation of particle interactions.
Fission	Break-up of a large nucleus into smaller ones.
Flavour	The generic name for the qualities that distinguish the various quarks: up, down, strange, charm, bottom, and (hypothesised) top. This concept also applies to leptons where it distinguishes electron from muon or tau and neutrino. Thus flavour includes electric charge and mass.
Fusion	Combination of small nuclei into larger ones.
Gamma ray	Photon. Very high energy electromagnetic radiation.
Gauge theories	A class of theories of particles and the forces which act on them of which quantum chromodynamics and the theory of the electroweak force are examples. The term 'gauge' meaning 'measure' was introduced by Hermann Weyl 50 years ago in connection with properties of electromagnetic theory and is used today mainly for historical reasons.

Generation	Leptons and quarks come two by two. Two leptons (such as electron and neutrino) and two quarks (such as up and down) form a generation. The first generation is $(e^-, \nu_e; u, d)$, the second is $(\mu^-, \nu_\mu; c, s)$ and the third $(\tau^-, \nu_\tau; t, b)$. The top quark (t) in the third generation is predicted by some theories but not yet seen.
GeV	A thousand million electron volts. Sometimes called BeV: B for billion (US variety).
Gluon	Carrier of interquark force. Plays a role in QCD analogous to that played by the photon in QED.
Hadron	Particle that can experience the strong force.
Higgs particle or **Higgs boson**	Particle whose existence is predicted by the electroweak theory (p. 120). It is believed that its existence is fundamentally connected with the massiveness of the W and Z bosons.
Ion	Atom carrying electrical charge due to its being stripped of electrons (positive ion) or having an excess of electrons (negative ion).
Intermediate vector boson	Generic name for W and Z bosons—the hypothesised carriers of the weak force.
J-meson	Name for the J or ψ meson, mass 3 GeV, compounded of a charmed quark and charmed antiquark. Its discovery in 1974 instigated a scientific revolution in the later half of the decade (Chapter 9).
Kaon (K-meson)	Variety of strange meson.
KeV	A thousand electron volts.
Lepton	Particle, such as electron, which does not experience the strong force.
Magic numbers	Many properties of nuclei vary periodically in a manner analogous to the periodic table of the elements. Especially stable nuclei occur when either the number of protons or neutrons equals the 'magic numbers', 2, 8, 20, 50, 82, 126.
Magnetic moment	Quantity that describes the reaction of a particle to the presence of a magnetic field.
Meson	Subnuclear particle consisting of quark and antiquark.
MeV	A million electron volts.
Molecule	Cluster of atoms.

Muon	Charged lepton. The analogue of electron in second **generation**.
Neutrino	Electrically neutral, massless lepton. There are three known varieties, one in each generation, associated with electron, muon, and tau leptons. Only takes part in weak interactions.
Neutron	Electrically uncharged baryon, partner of the proton in nuclei.
Neutral current	Weak interaction where no change takes place in the charges of the participants.
Nucleon	Generic name for neutrons and protons which are the constituents of a **nucleus**.
Nucleus	Dense centre of atoms, built from neutrons and protons. The latter give the nucleus a positive electrical charge by which electrons are attracted and atoms formed.
Orbits	Paths traversed by electrons in atoms.
Parity	The operation of studying a system or sequence of events reflected in a mirror.
Periodic table	Tabulation of the chemical elements which exhibits a pattern, in the regular recurrences of similar chemical and physical properties. Forerunner of the electron-orbiting-nucleus model of the atoms.
Photon	Carrier of the electromagnetic force.
Pion (or π-meson)	Lightest meson. Predicted by Yukawa, to explain the force binding the nucleus. Comes in three varieties distinguished by their electrical charges $+1$, 0, -1 labelled π^+, π^0, π^-.
Positron	Antielectron. Carries positive electric charge. Annihilation with electron produces energy and new varieties of hadrons and quarks.
Proton	Positively charged constitutent of the **nucleus** that gives it electrical charge. Built from (three) quarks.
Psi	A name by which the J or psi (ψ) meson is known.
QCD, QED	*Quantum chromodynamics* and *quantum electrodynamics* the theories describing interaction of colour and electric charges. The carriers are respectively gluons and photons.

Quark

Believed to be one of the fundamental constituents of matter. Distinguished from **leptons** by its possession of colour and fractional electrical charge.

Radioactivity

Spontaneous decay and transmutation of nuclei with emission of alpha, beta, or gamma radiation.

Spin

The intrinsic **angular momentum** possessed by a particle.

Strangeness

Property possessed by all matter containing a **strange quark**. This quark has charge $-\frac{1}{3}$ and partners the **charmed quark** in the second **generation**.

Strong interaction (force)

The strong attractive force that grips protons in the nucleus, overcoming their mutual electromagnetic repulsion. Now believed to be a remnant of a more powerful force between quarks—see **QCD**, **colour**.

SU(N)

Mathematical structure known as a 'group' that describes operations on N objects. Examples include SU(2) applied to the two quarks or two leptons in a generation and SU(3) applied to the three colours of quark. The three colours and two flavours have recently been combined to yield a set of five entities that can be described by a grand unified theory exploiting SU(5).

Symmetry

If a theory or process does not change when certain operations are performed on it, then we say that it possesses a symmetry with respect to those operations. For example a circle remains unchanged under rotation or reflection. It therefore has rotational and reflection symmetry.

Synchrotron

Modern form of particle accelerator.

Tau lepton

Negatively charged lepton in the third generation. A heavier analogue of **electron** and **muon**.

Thermal equilibrium

The particles of a gas (for example) are continuously in motion. The average speed of the particles is a measure of the temperature of the gas as a whole, but any given particle could have a speed that is much less or much more than the average. Thus there is a distribution of particle speeds within the gas. If the numbers of particles entering a given range of speeds (more precisely, velocities) exactly balances the number leaving, then the gas is said to be in thermal equilibrium.

Top
A flavour of quark, electrical charge $+\frac{2}{3}$ predicted by some theories to exist to complete the third generation of quarks by partnering the **bottom** quark.

Uncertainty principle
To measure both position and velocity simultaneously, requires two measurements. The act of performing the first measurement will disturb a particle and so create uncertainty in the second. Thus one cannot measure both position and velocity to perfect accuracy. The disturbance is so small that it can be ignored in the macroscopic world, but is dramatic for subatomic particles.

Unified theories
The attempts to unite the theories of the strong, electromagnetic, and weak forces of nature. Ultimately it is hoped that gravity will also be incorporated in this scheme.

Up quark
Lighest quark with charge $+\frac{2}{3}$. Partner of down quark in first generation. Constituent of proton and neutron.

Upsilon
Very massive (9 GeV) meson built from a bottom quark and bottom antiquark. Discovered in 1977, it is a member of the most massive family of particles known at present.

Van der Waal's force
Electromagnetic force between atoms or molecules due to imbalance between their positive and negatively charged constituents, even though the overall charge is zero.

Vector meson
A meson possessing one unit of spin—the same as the photon. Examples include ρ, ω, ϕ, ψ, Υ, which can all be produced directly by annihilation of an electron and positron.

Weak isospin
See page 115.

W boson
Massive electrically charged particle that carries the weak force in **charged current** weak interactions

Weak interaction (force)
One of the fundamental forces of nature. Most famous manifestation is in β-decay, also involved in some radioactive decays of nuclei, and neutrino interactions.

Weinberg angle (θ_W)   Parameter in the **electroweak** theory. Relates properties, such as mass, of the W and Z bosons and their interactions.

Weinberg–Salam model   A name for the **electroweak** theory.

X rays   Penetrating form of electromagnetic radiation (see Box 3.6 on page 29) discovered by Röntegen in 1895.

Z boson   Hypothesised massive electrically neutral particle that carries the weak force in **neutral current** interactions.

Index